Practical Agricultural Technology Book Series for Africa
Edited by Center of International cooperation Service, MARA

Practical Agricultural Technologies for Ethiopia (I)
埃塞俄比亚农业实用技术（上）

◎ Written by Xie Junhua, Liu Jianjun, et al.

China Agricultural Science and Technology Press

图书在版编目（CIP）数据

埃塞俄比亚农业实用技术.上/谢俊华等著.—北京：中国农业科学技术出版社，2020.12
（非洲农业实用技术丛书；1）
ISBN 978-7-5116-5107-5

Ⅰ.①埃… Ⅱ.①谢… Ⅲ.①农业技术—埃塞俄比亚 Ⅳ.①S

中国版本图书馆 CIP 数据核字（2020）第 247442 号

责任编辑　徐定娜　李　雪
责任校对　贾海霞

出 版 者	中国农业科学技术出版社
	北京市中关村南大街 12 号　邮编：100081
电　　话	（010）82105169（编辑室）（010）82109702（发行部）
	（010）82109709（读者服务部）
传　　真	（010）82109707
网　　址	http://www.castp.cn
发　　行	各地新华书店
印 刷 者	北京建宏印刷有限公司
开　　本	880 mm×1 230 mm　1/32
印　　张	7.5（全三册）
字　　数	345 千字（全三册）
版　　次	2020 年 12 月第 1 版　2020 年 12 月第 1 次印刷
定　　价	88.00 元（全三册）

◆◆◆ 版权所有·侵权必究 ◆◆◆

Editorial Board

Editor in chief:

Tong Yu'e

Associate editors in chief:

| Luo Ming | Lin Huifang | Hong Zhijie | Xu Ming |

Members:

| Wang Jing | Wei Liang | Fu Yan | Zhou Min |
| Yang Yang | Li Jun | Li Jing | |

Practical Agricultural Technologies for Ethiopia (I)

Xie JunHua	Liu Jianjun	Zhao Haizhi	Deng Zhengrui
Li Rongbiao	Li Shuhan	Liu Shouyun	Zheng Aibao
Zhu Zhenrong			

Contents

Year Round Management of Apple Orchard (Suitable for Deciduous Orchard) 1

Techniques of Intensive Wheat Cultivation in Ethiopia 13

Proposals for Whole-Process Mechanization of Main Crops Production in Ethiopia 20

Rice Nursery Seedling Preparation 25

Practical Manual for Cotton Specialty in Ethiopia 34

Tilapia Natural Reproduction and Farming Technology 40

Key Points of Tomato Cultivation Techniques for High Yield and Good Quality 49

Practical Technology on Low Cost Cultivation of Oyster Mushroom 55

Technical Guidance for Whole-Process Mechanization of Maize Production in Ethiopia 64

Year Round Management of Apple Orchard (Suitable for Deciduous Orchard)

1 In Spring (Germination, Flowering and Young Fruit Stage)

After the dormancy, with the temperature warm up, apple trees began to experience the process of germination, flowering and fruiting. The nutrients stored in the roots and branches are transporting to the shoots, so that the branches become soft and the buds begin to expand and germinate. The specific contents are as follows.

1.1 Fertilizing, Irrigating, Mulching and Moisture Conservation

Orchards that did not receive basal fertilizer in autumn and winter of last year should to Fertilizing immediately after thawing (but the effect is not as good as that in autumn). After fertilization, if there is irrigation condition, the orchard should to water once, and the soil of orchard should be hoed in time to keep the moisture. Fertilizing requirements: first, excavate the base fertilizer ditch to avoid damaging the root; second, the fertilizing position should be more than 30cm away from the trunk to avoid damaging the trunk. According to the weather conditions in spring, irrigate the orchard before and after flowering to ensure the quality of flowering and improve the fruiting rate.

1.2 Bud-notching and Bud-wiping

Bud-notching: before sprouting in spring, cut 1/4 to 1/3 of xylem at 0.5cm above (central main branch) or in front (oblique main branch and auxiliary supporting branch) to promote sprouting and branching.

Bud-wiping: after germination, wipe the buds on the back of the central main and auxiliary branches to avoid germination and branching, in order to wasting nutrition and disturbing the tree shape.

1.3 Making and Spraying Slime-sulfur Mixture

At present, slime-sulfur mixture is still a broad-spectrum, high-efficiency, long residual period, and low-cost bactericide and insecticide, especially the high concentration of spraying before and after the germination of fruit trees, which has a very significant effect on reducing the base of many diseases and insect pests, reducing the annual drug use and the cost of producing in whole year. Generally, it is better to spray 4–5 degree of slime-sulfur mixture mixed with 1/1,500 consistence of neutral washing powder on over 3 years old apple trees at germination stage.

The preparation method of slime-sulfur mixture: prepare 10 parts of sulfur powder, 7 parts of quicklime, 60 parts of water. Take a small amount of hot water to mix the sulfur powder into paste and put it into the pot to boil. Slowly put the quicklime into the pot, increase the fire power, and continuously stir until the lime block is put into the pot finish and boil for another 45 minutes (the first 15 minutes use the strong fire, the next 30 minutes use the gentle fire) until the liquid is in the color of brown. Measure the accurate concentration with Baume specific gravity after cooling. Generally, it is better to spray 4–5 degree of slime-sulfur mixture mixed with 1/1,500 consistence of neutral washing powder on over 3 years old apple trees at germination stage.

1.4 The Second Time of Pruning

There are 3 sides of target: The first is to delay the winter pruning of trees that are too prosperous and do not bear fruit at the right age until they have sprouted, so as to ease the tree's vigor. The second is to delay the pruning of other branches of trees that are growing more vigorously, except for the winter pruning of bone trunk branches, so as to ease the branch's vigor. The third is to enter the fruiting period, according to the target yield of trees. If the flower amount is too much, a part of medium and long flowering branches can be cut to short, a bunch of flower branches can be cut a part of it, or weak and short flowering branches can be thinned, In order to reduce the number of flowers and increase the number of preparatory branches of fruiting.

1.5 Spraying Foliar Fertilizer and Releasing Bees

In time to spray 0.3% Borax + 0.1% Urea + 1% Sucrose aqueous solution once at the initial flowering period (5% central flowers are blooming); spray again when entering the full bloom stage 0.3% Borax added with 1% honey water solution to facilitate pollination and fruit setting. A box of bees placed within 500 meters from the orchard can ensure pollination of about 10 mu (about 6,670 m^2) and increase the fruiting rate by 30%–50%.

1.6 Flower and Fruit Thinning

The purpose of flower and fruit thinning is to improve fruit quality (internal and external quality), reduce unnecessary consumption of nutrients and improve economic benefits. At the same time, it is beneficial to preserve the tree vigor and maintain the yield and quality of the next year.

2 In Summer (Fruit Growth and Expansion Period)

2.1 Tree Management (Summer Pruning)

(1) Bud-wiping: Timely remove the buds on the base of the tree and the back of the branches will help to reduce the nutrient consumption, reduce the workload of pruning in winter and optimize the structure of the tree.

(2) Pick off the top-buds and twist the soft branch: It can effectively control the vigorous growth and promote the formation of flower bud by picking or twisting the competitive branches on the head of main branch and the erect vigorous branch on the back of branches. If there is space, leave 3–4 leaves to pick off the center buds to control the growth. When the new shoot grows to about 20 cm, to twists it at 5 cm of its base.

(3) Pull the branches: Pulling branch is one of the most important technical measures for tree culture and tree improvement. According to the requirements of the tree structure, the first is to control the angle of the pulling, the main branch is pulling to 70° to 90° angle, the twigs and fruiting branches on the main branch are drooping, pulling to 100° to 110° angle. The second is to adjust the direction of branches, arrange

the branches evenly, occupy space reasonably, without overlapping, joint and crowding.

(4) Bud-notching: At the above about 0.5 cm of the bud cut deep into the xylem can improve germinate of short branches and overcome baldness.

(5) Thinning branch: Remove the peripheral multiple new shoots, cut the new shoots from the pruning position, empty cut the long branches in the inside of the tree, and over cut dense upright branches on the back of the main branches. After thinning, the branches in the same direction are keep about 30 cm apart and cannot thinning away all branches.

(6) Twisting branch: For 1–2 year-old long branches, it is better to fold down the branches by hand, but not break off. Once every 10 cm between, to change the branch potential, weaken the apical dominance, and promote germinate of short fruiting branches.

2.2 Fertilizing in Summer (Topdressing)

The growth of apple trees is vigorous in summer, and it is also a period of rapid fruit expansion. The nutrient consumption is very large. Timely and reasonable application of top dressing is conducive to maintaining the balance of nutritional growth and reproductive growth, promoting fruit quality, and avoiding the phenomenon of "big-small wave years" caused by nutritional imbalance. Pay attention to the time, type and method of fertilization.

(1) Topdressing time: it is better to topdressing in the period of fruit expansion and flower bud differentiation.

(2) Types of fertilization using: The combination of N, P and K is the best, which accounted for 10%, 20% and 80% of the annual fertilization

respectively.

(3) Amount of fertilizer applied: There are several methods to determine the amount of fertilizer application, such as empirical fertilization (conventional fertilization), leaf analysis and fertilization according to the amount of results. In the production of fruit trees, the method of determining the amount of fertilizer is often used. Generally, it needs to absorb N 0.5 kg, Phosphorus (P_2O_5) 0.3 kg and Potassium (K_2O) 0.6 kg to produce 100 kg apples.

The amount of fertilizer application should be determined according to the age of trees, tree potential, load of producing and soil conditions. The mature orchard needs about 6–10 kg of pure Nitrogen, 2–5 kg of pure Phosphorus and 8–15 kg of pure Potassium annually. It is 20% of the annual amount of fertilizer application in this time of top dressing.

(4) Methods of fertilization: Most of them adopt the methods of radial furrow and hole-shape fertilization. The integrated technology of water and fertilizer can also adopt to use, the fertilizer is better to choose compound microbial, the high phosphorus water-soluble fertilizer, and the high potassium water-soluble fertilizer can be selected as the fertilizer.

2.3 Pest, Disease and Parasite Control in Summer

Summer is a long time for apple trees to flourish and grow rapidly. The quality of photosynthesis plays a decisive role in the growth of trees and fruits. Therefore, the prevention and control of diseases and insect pests will help to maintain the healthy photosynthetic leaf amount, maintain the balance of tree growth, and improve the quality and commodity rate of fruits. Mites, Carnivores, Brown spot and Black spot were mainly controlled.

The principle of agricultural chemicals using. Select the high-efficient insecticides and less destructive to natural enemies for diseases and insect. Select the insecticides with low coagulability for fruit surface and no damage to young fruit, and do not use the chemicals easy to cause drug damage such as mancozeb etc, and select the high-quality insecticides and bactericides with no residue for fruit and no pollution to the environment.

Investigate and clean the parasite of fruit trees

3 In Autumn (After Harvest Period)

After fruit ripening and harvesting, fruit trees gradually turn from rapid growth to nutrient accumulation and storage, which stores nutrients for growth, flowering and fruit in the next year. The main contents of autumn management of apple orchard include pruning in autumn, applying base fertilizer in autumn, picking leaves and turning fruits, protecting fruit surface, fruits harvesting, etc.

3.1 Pick Leaves and Turn Fruit

First is pick off the shading and the leaves attached to the fruit within 15–20 cm around the fruit. Then after 5–6 days remove the shading leaves, thin leaves, yellow leaves and old leaves around the fruit, and then remove the part of leaves that affect the light transmission of autumn shoots and middle and upper branches, and try to retain the functional leaves, so as not to affect the photosynthetic efficiency. When picking leaves, the petioles must be kept, and the number of leaves to be picked should not exceed about 20%–30% of the whole tree. Gently rotate the fruit to make its back and shade turn to the sun. Do not use too much force to avoid twisting the fruit. It is better to rotate the fruit 2–3 times. It also can combined with the laying of reflective film to accelerate the fruit coloring and even coloring.

3.2 Fruit Surface Protection

Generally, the fungicides such as Carbendazim and zincic are sprayed on the fruit surface of non-bagged fruit. After bagging, the skin of the fruit is delicate, and it is very easy to infected with red spot disease. The pores increase, resulting in cracks. In addition, calcium deficiency, calcium deficiency and other diseases are very easy to occur, and small cracks appear on the fruit surface, reducing the storage and commodity performance of the fruit. Therefore, spray 1–2 times of fungicide and calcium fertilizer after bag removal.

3.3 Fruit Harvesting

Generally, commercial fruits should colored by natural sunshine fully and evenly, the best harvest time is when they are 70%–80% of maturity. During harvesting, should pay attention to the preservation of fruit stalks, handle carefully, and store in grades. At the same time, pay attention to the harvesting weather, generally do not in rainy days and the early morning with high air humidity.

3.4 After Harvest Pruning

The objects of thinning are cross branches, parallel branches, overlapping branches and competitive branches. At the same time, combined with twirling, kneading and pulling branches to carry out tree management.

3.5 Base Fertilizer Application in Autumn

The best time to apply fertilizer is September, no later than the first ten days of October at the latest. The basic fertilizer is mainly mature farm manure, including chicken manure, sheep manure, biogas residue, etc. the fertilizer effect of crop straw is low, and the amount can be increased up properly. Appropriate chemical fertilizer, especially P-K fertilizer, such as farm manure. The common fertilization methods of base fertilizer are strip furrow, ring furrow and radial furrow.

4 In Winter (Dormancy Period of Fruit Trees)

After defoliation, the growth and sap flow of apple trees almost stopped. A good winter management plays an important role in apple orchard management, such as strengthening the tree, delaying tree senescence, eliminating pests, adjusting the size of the year, and improving the quality of fruit in the next year. The main contents of winter management include winter pruning, tree protection (pest control), watering, etc.

4.1　Dormancy Period Pruning

The suitable period of apple tree pruning in winter should be controlled flexibly according to the variety, age and vigor. If pruning is too early, the cutting mouth will easily damaged by freezing, forming dry piles; if pruning is too late, the tree will weaken and affect the yield. The general principle of pruning time selection is to start pruning after the tree is completely dormant, and finish pruning before the sap starts flowing.

The general principle of pruning is: young trees are mainly used for shaping, while adult trees are mainly used for ventilation and light transmission, which is conducive to fruit setting and improving the quality of fruit, focusing on the next year and taking into account the long term.

4.2　Protection of Fruit Trees (Pest Control)

In winter, it is the period of dormant overwintering of fruit diseases and insect pests. The location of dormant insect pests is relatively fix to hide, which is a good time to kill the diseases and insect pests and effectively reduce the occurrence of diseases and insect pests in the next year.

Dormancy period pruning

4.2.1 Clean the Orchard

When more than 90% of the fruit trees have fallen leaves, it is the time to clean up the orchard. All the items that may provide overwintering places for diseases and insect pests such as apple moth, such as dead branches and leaves, weeds, diseased branches, fallen fruits and waste chemical fertilizer bags, etc., shall be burned in a centralized way to eliminate the overwintering diseases and insect sources in large quantities.

4.2.2 Bark Scraping

During the dormancy period of fruit trees in winter and the time before germination in early spring. Scraping off the rough bark and warped bark below the main branches of fruit trees can eliminate the overwintering insect sources, such as the apple moth, the red spider, the little leaf roller moth, etc. and detect the branch and trunk diseases, such as the apple tree rot, the branch and trunk ring rot and the dry rot. When scraping, do not damage the xylem of the fruit tree. Take the scraped tree out of the orchard and burn it intensively. After peeling, white the dried fruit branches in time.

4.2.3 Deep Turning Soil

After clearing the orchard and before the soil is frozen, turn 20–30 cm deep under the canopy around the fruit tree, and then irrigate water to change the environmental conditions of the soil and destroy the overwintering place of the pests. It can reduce the sources of overwintering insects, such as apple moth, peach moth, beetle, etc.

4.2.4 White Coating

After the fruit tree leaves and before the soil is frozen, use lime white coating agent to white the trunk and branches of the fruit tree. The height of white coating is generally 60–80 cm. The proportion of white coating agent is generally: 10 parts of quicklime, 2 parts of stone sulfur mixture, 2 parts of salt, 2 parts of clay and 40 parts of water, stir to make it mixed evenly. It should be painted again in early spring. It can prevent the occurrence of the fruit tree's daily burning disease and freezing injury, and can also eliminate the pests and pathogens such as the larvae of the apple moth that overwinter on the tree trunk.

4.3 Watering in Period of Dormancy

In period of dormancy, irrigate the whole orchard or single plant once, but it must be fully irrigation. In this way, the water holding capacity of soil can be increased, the decomposition of soil nutrients can be promoted, and pests can be eliminated.

Techniques of Intensive Wheat Cultivation in Ethiopia

Wheat is planted in high land and mid highland in Ethiopia, which is one of Ethiopia's most important cereal crops by production. About 4.8 million farmers produce close to 4.642 million tons of wheat across 1.69 million hectares of land (2018), the average yield is 2,747 kg/ha. According to the land coverage, wheat is the fourth crop in Ethiopia, after Teff, maize and sorghum, but the wheat production can not meet the increasing demand in Ethiopia.

Wheat field in one of the major producing areas (ARSI), Ethiopia

The national level evidence shows the increases in total fertilizer imports as well as in the applied volume of fertilizer. However, the rates still remain low at 29 kg per hectare for arable and permanent cropland

and 37 kg per ha for grain production compared to 200 kg per ha that is generally recommended as optimal for crop production in the country (World Bank, 2012).

Incidence of the three types of the rust diseases (yellow rust, leaf and stem rust) has been on the rise in the past few years in Ethiopia.

Due to the current conditions in the field of cultivation, comprehensive cultivation techniques should be followed to increase wheat yield and improve the quality of wheat.

1 Location Selection

1.1 Soil Quality

To get the best yield, it's essential to choose topography, soil permeability, and easy to cultivate and soft fertile sandy loam or loam. Soil and water management practices should be intensified to promote water (in case of moisture deficiency) and soil conservation so as to increase wheat productivity. On the other hand, drainage of excess water should be ensured in water-logged vertisol areas. In areas of acidic soils, the existing scheme of soil treatment with lime should be intensified. Black soil and red soil are all suited for wheat planting. The optimum pH is 5.5–7.5.

1.2 Temperature

The optimum temperature for wheat growth is 15–25 degree Celsius.

1.3 Altitude

The wheat grows between 1,500–3,200 meters above the sea level, the optimum altitude is 1,800–2,800 meters above the sea level.

1.4 Rainfall

The needed annual rainfall for wheat planting is 500–1,200 mm, but 800–1,000 mm rainfall is suitable.

2 Crop Rotation

Crop rotation should be promoted in the traditional wheat dominated farming system to minimize build-up of grass weed species due to mono-cropping. The planned rotation sequence may be at least for two or three years, beans are the best preceding crops. Potato, corn, cotton and canola are also good candidates.

3 Land Preparation

3.1 Extensive Preparation of Soil for Full Seedlings

The whole season is divided into two parts in Ethiopia, rainy season and dry season. Generally when small rainy season begins, it's better to till the land with the first two soaking rains. If there is a longer interval before sowing, plowing can be repeated several times to eliminate weeds. Before sowing seeds, at least two times land plowing will be followed. Early farming can't be conducted for the dry land, but delayed plowing tillage will be difficult for the wet sticky soil. It's better to arrange the tillage time carefully and finish the work when the soil moisture is medium.

After soil preparing, the soil should be loose enough, the aeration condition is improved and the surface becomes even, no huge crop residues and stubbles, the soil is suitable for sowing.

3.2 Applying Enough Basal Fertilizer

So far DAP, Urea and NPS are widely used in Ethiopia. There is no manure using habit in most parts of Ethiopia, so it's necessary to actively publicize and promote the farmers to use manure to improve soil fertility thus to improve the yield potential.

The Recommended Fertilizers Application for Wheat

Soil types	Fertilizer types	Application amount(kg/ha)	Time of application
Black soil/ clay soil	NPS	100	Seeds Sowing time
	Urea	250	1/3 at seeds sowing time, 2/3 applied 30–35 days after sowing seeds and the weed is properly weeded.
Red soil	NPS	100	Seeds Sowing time
	Urea	200	1/3 at seeds sowing time, 2/3 applied 30–35 days after sowing seeds and the weed is properly weeded.

4 Variety Selection and Seed Treatment

4.1 Variety Selection

Seed is one of the most important inputs for improving productivity of crops, particularly in resource-constrained SHFs. The main wheat varieties cultivated in recent years are Hulluka, Ogolcho, Shorima, Hoggana, Danda'a, Kakabat etc.

4.2 Seed Treatment

Selected wheat seeds are treated by the ratio 1:10 phoxim solution, 100 g phoxim is for treatment of 10 kg wheat seeds, during the process, the treated seeds should be bored pile for 3 to 4 hours, when the treated seeds are dry, every 10 kg seeds are mixed with 2 g 15% triadimefon. Till now Seed coating technology has not been used in this country due to some reasons, so it is necessary to introduce this technology to Ethiopia.

5 Sowing

5.1 Optimum Sowing Dates

The main wheat growing areas of Ethiopia are the highlands of the northwest to Addis Ababa, central Ethiopia and south-eastern to Addis Ababa. The optimum sowing dates are in the medium of June to the early July.

5.2 Seeding Rate and Planting Depth

Usually 100–125 kg good quality seeds per hectare is recommended in Ethiopia and sowing depth is about 3 to 4 cm.

5.3 Sowing Methods

There are many ways for wheat sowing, three commonly used introduced here.

5.3.1 Mechanical Sowing

Mechanical sowing has the advantages as uniform sowing, easy to keep soil moisture, high efficiency and straight lines. The distance between the rows is 20 cm.

5.3.2 Manual Broadcast Sowing

The plough pulled by the livestock plows, trenches, then broadcasts seeds by hand and covers the seeds.

5.3.3 Manual Row Planting

The plough pulled by the livestock plows, trenches, then delivers seeds by hand in rows. Unusually the distance between the rows is 20–25 cm, as long as the suitable row space is controlled, the sowing depth is guaranteed, because wide row space usually causes the deep sowing that is not good for germination.

5.3.4 Sowing Requirements

No matter what the sowing method is used, the basic requirement should be met, such as delivering seed evenly, no missing and discontinuous ridge, the same depth, etc.

6 Field Management

6.1 Early Soil Loosing to Promote Healthy Seedling

Early soil Loosing is conducted when the seedling is 4 cm tall (about 2–4 true leaves).

6.2 Water Management

In most parts of Ethiopia, wheat is rain-fed crop. If it is in the dry condition and possible, wheat should be irrigated at the seeding stage (2 true leaves), jointing stage, booting stage and grain filling stage.

6.3 Apply Top Dressing Fertilizer to Enhance Wheat Yield

In high level productivity land, the top dressing fertilizer is applied. 2/3 of Urea fertilizer applied 30–35 days after sowing seeds and the weed is properly weeded. At the bloom stage, topdressing fertilizer 1% urea solution can be sprayed, which is determined by the crop nutrient requirement and soil supply amount.

6.4 Subsoil Early and Frequently

At the seeding stage, if soil humidity is high, deep hoe to 4–5 cm to scatter the soil moisture. If it is severe drought, dig shallow but repeat several times to maintain soil moisture. In case of termites, early dressing applying and more subsoil should be taken to reduce the production loss. That is to say, hand weeding should also be practiced timely.

6.5 Spraying Chemicals for Disease Control

Rust diseases (yellow rust, leaf and stem rust) are the main disease affect the wheat during the jointing stage. Tilt 250 EC, Bayleton 25 WP, Bumper 25 EC, Noble 25 WP, Opera TM Max, Rex® Duo, Thiram Granuflo 80 WP are the registered chemical in Ethiopia. The most effective and widely used is Tilt 250 EC, which is sprayed during jointing stage to grain filling stage.

6.6 Harvest on Time

When the wheat hulls become yellow and the grains become harden, it's the right time for harvesting. Carefully take measures to prevent from birds pecking in case of the yield loss. In the main wheat produced areas such as Arsi, Bale, Goba, wheat is harvested by combine harvester. In other areas, it is harvested by hand sickle cutting, then the seeds removed by animals stepping or small scale thresher, at last store the grain in dry and cool place.

Proposals for Whole-Process Mechanization of Main Crops Production in Ethiopia

At present, the level of mechanization in agricultural production is without balance in Ethiopia. In contrast, there is a certain foundation of mechanization in the fields of arable land and wheat harvest, but the lower mechanization still in many areas of agricultural production, so the whole-process mechanization of main crops Production should be promoted which is a practical way to raise the level of agricultural mechanization in Ethiopia in a planned and targeted way.

Whole-process mechanization of main crops Production aims at, surrounding agricultural development requirements, to promote advanced and applicable agricultural mechanization technologies and equipment in the whole process of the production of main crops, especially to improve the mechanization level in farming, sowing, plant protection, harvesting and straw processing, cultivate agricultural machinery service markets, explore mechanized production models, improve agricultural mechanization infrastructure, create demonstration areas, and build a new mechanism for the development of agricultural mechanization.

1 Content

Crop Types: Teff, rice, corn, wheat, potato, cotton, soybean, sugarcane and other main crops.

Production Links: To improve the level of mechanization of the main links such as tillage, planting, plant protection, harvesting and straw process.

Main Direction: The First is to improve the mechanization of the production of major food crops. The key is to improve the level of mechanized operations in land preparation, sowing of all crops, harvesting of corn, potato, soybean and teff, solve the problem of weak links in the development of mechanization. The second is to solve the problem of mechanization in the whole process of the production of major cash crops, and focus on demonstrating and popularizing agricultural mechanization technology in key links such as mechanized planting and harvesting of cotton, sugarcane and other cash crops.

According to the advantageous producing areas, planting patterns and mechanization development requirements of main crops in Ethiopia, It will explore a series models of whole-process mechanization of main crops production, establish the major content of promoting mechanization of the whole process of each main crop production, establish a number of demonstration areas that will take the lead in basically realizing whole-process mechanization of production by crop and region.

2 Implementation

(1) According to the economic conditions, production scale, mechanization level and other factors in different regions, the suitable promotion mode and agricultural machinery are selected to form a whole-process mechanized production mode with regional characteristics.

(2) It will give priority to promoting whole-process mechanization

in major grain-producing areas, and develop demonstration areas and sites for modern agriculture, which will be developed step by step from one unit and popularize it in the whole area.

(3) It will develop farmland infrastructure and promote the integration of agricultural machinery and agronomy, and the integration of agricultural mechanization and information technology.

(4) The government shall support and guide the development of socialized service organizations for agricultural machinery and large-scale operators of agricultural production. attract the participation of agricultural machinery manufacturers and departments of agricultural scientific research, extension and education to promote the Whole-process mechanization of main crops.

(5) Formulate medium-and long-term development schedules, make clear phased development goals, carry out regular work summary, timely identify problems, and adjust the focus of work at different stages.

3 Keypoint

(1) The government should strengthen its guiding role. and establish a sound quality management system to ensure the advancement and applicability of agricultural machinery in the Ethiopian market and prevent inferior agricultural machinery products from entering the Ethiopian market. Establish a complete standardization system for agricultural machinery, formulate technical specifications and technical standards for mechanical operations in each link of main crop production, and guide the standardization operations in each link of crop production; strengthening training and technical guidance.

(2) Developing social services for agricultural machinery, cultivate diversified socialized service organizations for agricultural machinery actively and promote cross-regional operations, order operations, trusteeship services, leasing and other agricultural machinery socialized services vigorously to Increase mechanization of crop production. We will guide nongovernmental investment in agricultural machinery operations, and promote their marketization, specialization, scale and industrialization in agricultural machinery service. We will foster a number of agricultural machinery cooperatives to hold demonstration activities.

(3) Establishing demonstration areas for whole-process mechanization of main grain crops and explore regional whole-process mechanization production models, encouraging governments at all levels to increase investment in whole-process mechanization to carry out experiments and demonstrations on whole-process mechanization actively, exploring and summarizing technical paths, technical models, mechanical equipment, operating procedures and service methods for whole-process mechanization, and accelerate the promotion of advanced and applicable agricultural mechanization technologies. Exploring ways to mechanize the production of major grain crops in different regions and set up a number of whole-process mechanization demonstration projects. By setting up replicable and popularize models, the mechanization level of the whole process of main crops production in the surrounding areas can be continuously improved.

(4) Strengthening agricultural mechanization infrastructure and improving conditions for the development of whole-process mechanization. Strengthen the construction of standardized farmland, actively

promote the construction of farmland water conservancy infrastructure and land consolidation; promote the construction of machinery informatization, build a national unified monitoring and service platform for dynamic information of agricultural machinery operations, timely collect and release information on supply and demand of agricultural machinery operations, and cultivate and standardize the market for agricultural machinery operations; we will support agricultural machinery cooperatives in building warehouses and sheds for agricultural machinery and tools, and strengthen the construction of roads and repair outlets for agricultural machinery, so as to solve the problems of "difficult housing, difficult roads and difficult maintenance" of agricultural machinery.

4 Objective

It will aim to establish 100 whole-process mechanization demonstration areas in Ethiopia by 2025. Among them, teff, wheat, corn and other major grain crops have formed a mature whole-process mechanization development model. The overall mechanization level of farming and harvesting reached more than 30%, and the mechanization level of mechanized plant protection, prevention and control, and mechanized straw treatment was greatly improved. By 2030, about 500 demonstration areas will be built in Ethiopia, which will take the lead in basically realizing the whole-process mechanization of production. The comprehensive mechanization level of farming and harvesting of main grain crops in Ethiopia has reached more than 50%.

Rice Nursery Seedling Preparation

Rice seedling preparation is a key stage in rice production, an important measure to capture high yields of rice, and also a technical difficulty or challenge in rice production.

According to different water management methods, rice seedling raising methods are divided into: water seedling raising, wet seedling raising, dry seedling raising, two-stage seedling raising, plastic floppy seedling raising, etc. Combining with the actual situation in Ethiopia, this article takes water seedling raising method as an example to introduce the main technical points of rice seedling raising technologies.

We know that the most important environmental factors affecting plant growth are: temperature, humidity, radiant energy, mineral composition, soil permeability and soil structure, soil pH, chemical composition, mineral nutrient supply, unrestricted substance composition, etc. Each factor can become a constraint for plant growth. None of these environmental factors affect the growth and development of plants in isolation, such as the contradictory relationship between soil moisture and air content. The following are the four most important factors. In order to facilitate memory, I summarize it as WALT:

• Water (Humidity): There is no life without water.
• Air: plants also breathe, called respiration.
• Light: photosynthesis.

• Temperature: Photosynthesis, respiration, evaporation-water evaporation, water and fertilizer absorption are all directly related to temperature.

In fact, all the technical measures we take in the process of raising seedlings are constantly adjusting these environmental conditions to achieve the purpose of cultivating strong seedlings. Therefore, raising seedlings is a process of constantly adjusting WALT.

1 Seed Preparation

No matter what method is used to raise seedlings, the seed treatment before sowing must be done.

1.1 Variety Selection

According to local conditions, choose high-quality, high-yield, and highly resistant varieties that are authenticated and suitable for local cultivation.

1.2 Sun Drying

The purpose is to make the seeds germinate and dry, so that the seeds can soak up and absorb enough water; use the heat and short light waves of the sun to distribute the moisture and carbon dioxide generated by the seeds during storage, and at the same time enhance the permeability and water absorption of the rice seeds and promote the activation of enzymes, thereby improving The germination rate and germination potential are neat and consistent. To sun drying seed for 1 day before soaking.

1.3 Seed Selection

Seeds with full grains and high germination rate are high-quality seeds. Selection of seeds can help to deflate and stay full, reduce the difference in quality between seeds, make the seeds germinate neatly, and strengthen the seedlings. There are many methods for seed selection, including fan selection, water selection, brine selection, yellow mud water selection, etc. Generally, the selection is performed with clear water, and the seeds are released for about half an hour to remove the floating grain and diseased grains, Weeds, foreign bodies, etc.

When using the salt water method for seed selection, the specific gravity of the water solution is 1.06 to 1.10 (a fresh egg could be placed in the salt water, and the surface area of the egg is about 1 coin size). After the salt water selection is done, remove the grains, weeds, foreign objects, etc. floating on the surface, remove the seeds, wash the seeds with clean water, removing the salt outside the chaff to prevent the low germination, and dry it after washing or directly seed soaking.

1.4 Disinfection

Many diseases that occur during the growth of rice are transmitted through seeds, such as rice blast, verticillium wilt, bacterial leaf blight, flax leaf spot, and dry sharp nematode. Therefore, after seed selection, it is necessary to carry out disinfection treatment in combination with seed soaking. While immersing seeds, combine using chemical *seed immersion spirit, TCCA, prochloraz*, etc. to disinfect the seeds. One of the methods is to use 25% *prochloraz emulsifiable concentrate* 2,000–3,000 times solution (that is, a 2 ml *promethamine emulsifiable concentrate* mixed with water 4–6 kg, seed soaking 3–4 kg), intermittent seed soaking, hybrid rice seed soaking for 24 hours, the conventional rice seeds are soaked for 36 hours no need pre-soaking, and no need rinsing after soaking, they can be directly germinated and seeded. The total seed soaking time, including disinfection time: 48–72 hours for early rice conventional seeds, 36–48 hours for hybrid seeds, intermittent soaking, the seeds should absorb about 35% of the moisture.

1.5 Seed Soaking

The length of rice soaking is related to the water temperature of soaking. Generally, seed soaking for 5 to 6 days at a water temperature of 15 degrees, and soak for 7 to 8 days at a water temperature of 10 degrees. The standard for absorbing sufficient moisture of rice seeds is that the seeds should absorb 35% of the adequate moisture, the chaff is transparent, the rice grains are visible in white, and the rice grains are easy to break without noise.

1.6 Accelerating Germination

The main technical requirements for germination are "fast, orderly, uniform and strong". Fast means that the buds are urged within 2 days; Orderly means that the germination potential is required to reach more than 85%; Uniform means that the buds are neat and uniform in length; Strong means the young buds are strong, the ratio of root and bud length is appropriate, and the colour is bright white, the smell is fragrant, no alcohol. According to the main process and characteristics of seed growth and germination, germination can be divided into three stages: high temperature breaks the chest, moderate temperature for germination and spreading for cold domestication.

1.6.1 High Temperature Breaks the Chest

When the rice seed is piled up until the embryo breaks through the chaff to reveal white, it is called the chest-breaking stage. After the seeds have absorbed enough water, the appropriate temperature is the main condition for fast and neat chest breaking. Within the upper limit of 38 degrees Celsius, the higher the temperature, the more vigorous the physiological activities of the seeds, the faster and neat the chest breaking, otherwise, the chest breaking is slow, and untidy. Generally, after the temperature of the piled grains rises, the temperature of the upper and lower parts of the grain piles must be consistent, and if necessary, the piles should be turned over to make the heat between the rice seeds uniform, and promote the uniformity of the broken chest. Before germination, rinse and drain the soaked grains, then soak them in warm water (50–55 ℃) for 5–10 minutes, then drain and put on the pile, keeping the temperature of the heap 32–35 ℃. After 15 to 18 hours, it begins to show white.

1.6.2 Moderate Temperature for Germination

The germination stage is when the rice seeds are broken from the chest until the elongation of the young shoots reaches the sowing requirements. The standard for budding of water-raising seedlings is: the root length is equal to the length of the rice, and the bud length is half the length of the rice. According to the principle of "dry long root, wet long bud", the length of the bud root is mainly controlled at this stage by adjusting water to achieve this, at the same time, the temperature of the grain heap should be adjusted in time. Keep the temperature of the pile at the germination stage at 25 to 30 degrees to ensure the coordinated growth of roots and shoots and thick roots. After breaking the chest and exposing the white, the grains are turned over to dissipate heat, and drenched with warm water to maintain the temperature of the heaps at 28–32 ℃. After the roots are properly poured, warm water of about 25 ℃ is poured to keep the grain pile moist and promote the growth of young shoots.

1.6.3 Spreading for Cold Domestication

In order to enhance the adaptability of grain buds to the external environment after sowing and improve the uniformity of germination, cold sprouting should also be spread in the late stage of germination. Generally, after the germination of grain buds reaches the standard, it should be placed indoors and allowed to cool for 4 to 6 hours, and the seed moisture is suitable for sowing without touching your hands.

2 Nursery Preparation

2.1 Selection of Rice Seedling Nursery Field

It is required to choose a field as a seedling field with leeward to the sun, loose soil, high fertility, convenient drainage and irrigation, no pollution, less weeds, and convenient transportation. The ratio of seedling field to transplanting rice field is 1:8 to 1:10, and the seedling field is prepared 1:20 to 1:25 by using a floppy disk to raise seedlings and mechanical transplantation seedling is 1:80.

2.2 Tillage of Rice Nursery Fields

Plough the nursery field 30 days before sowing, combine levelling field 3 to 5 days before sowing, supply 25% compound fertilizer 380–450 kg per hectare of rice field as base fertilizer for seedbed, comb rake 2 to 3 times during fertilization to ensure uniformity does not hurt buds.

2.3 Size of Nursery Beds

The surface of seedling nursery field should be divided to width of 1.5 meters, the width of the furrow is 0.3 meters, the depth of the furrow is 0.15 meters, and the length of the nursery is not more than 10 meters. It is better to form the seedling beds in the direction of north and south, and open the waist groove and the surrounding groove to facilitate irrigation and drainage.

3 Seedling Nursery Management

3.1 Bud Stage (Before the First Leaf is Unfolded)

Take root and set up seedlings, generally only irrigate in the ditch, there is no water layer on the border surface, only keep the border surface moist, pay attention to protect the bud with water.

3.2 Young Seedling Stage (1–3 Leaves)

Shallow irrigation, when 1 and a half leaf bud for weaning fertilizer, 3 leaves for weaning fertilizer, 1 time fertilizer per leaf, 45–75 kg urea/ ha per time.

3.3 Seedling Stage (3 Leaves to Transplanting)

Water management: shallow irrigation; fertilizer: at 4 to 5 leaves age applying relay fertilizer, 45–75 kg urea/hectare, 5 to 7 days before transplanting, applying 45–120 kg urea/ hectares as dowry fertilizer.

4 Prevention of Diseases and Insect Pests

According to the occurrence of diseases and insect pests, prevention and control work should be done in the seedling stage. At the same time, weeds should be removed to ensure the purity of the seedlings. At the beginning, special attention should be paid to prevent rats and bird damage.

Rice seedling nursery, took in Kenema, Sierra Lione

Transplanted rice farm, took in Gambella, Ethiopia

Sun drying treatment for rice seed, took in Kalambeza, Namibia

Rice Nursery Seedling Preparation

Nursery bed field preparation, took in Kenema, Sierra Lione

Plastic floppy seedling raising, took in Kalambeza, Namibia

The seedling is reaching the age for transplanting, took in Kenema, Sierra Lione

Transplanting, took in Kalambeza, Namibia

Practical Manual for Cotton Specialty in Ethiopia

1　Cotton Plant Structure Adjustment Techniques

Is through manual or mechanical operation, remove part of cotton plant organs, adjust cotton plant morphology and structure, so as to achieve the purpose of increasing cotton yield and quality.

1.1　Theoretical Basis

Rational pruning can regulate nutrient growth and reproductive growth, regulate the transportation and distribution direction of nutrients in plants, reduce the unnecessary consumption of nutrients, and improve the yield and quality of cotton.

For cotton fields with excessive vegetative growth, pruning can improve the field microclimate, increase ventilation and light transmission. Increase the utilization rate of light energy, prevent the excessive growth, make the square and boll get enough nutrition, reduce the square and boll fall off and reduce rot boll, increase early-stage boll forming, promote precocity and increase yield and quality.

Plant structure adjustment can increase the permeability of spraying, also can induce cotton plant resistance, reduce the occurrence of pests and diseases, can improve the yield and quality.

1.2 Field Operations

The adjustment of cotton plant structure mainly includes: Topping, trimming, vegetative branch removal, redundant bud removing, old leaves and diseased leaves removal, empty branches and diseased branches removal etc.. In the actual operation process, according to the cotton growth situation, labor force situation and other actual conditions for all operations, or selective operations.

1.2.1 Topping

In order to control the growth of plant height and invalid fruit branches, the apical bud and one leaf at the top of the main stem of cotton plant was removed during the period of initial flowering period to flowering period.

Topping time: The topping time should be determined according to soil fertility, planting density and growth potential, generally 70–80 days before cotton harvesting. Timely topping can control the height of cotton plant, reduce unnecessary nutrient consumption, and make the available nutrients can be better transported to the square and boll, so as to increase the fruit setting rate and boll weight, which is beneficial to increase production and precocity.

1.2.2 Side Branch Trimming

In the case of excessive vegetative growth of cotton field, the terminal bud on the fruit branch was removed at the right time to control the growth of invalid fruit nodes, regulate the distribution of nutrients, improve the ventilation and light transmission conditions, increase the boll weight, promote precocity, control lateral growth, and facilitate the operation of cotton fields.

Generally, before the topping operation the lower part of the fruit branch on the main stem were trimmed, after the topping operation, the upper part of the fruit branch were trimmed.

1.2.3 Vegetative Branch Removal

There are 2–4 vegetative branches every cotton plant generally, bearing in the lower part of the main stem 3–7 note. Leaf branches grow fast, consume more nutrients, inhibit the growth of fruit branches and affect the development of square and boll. At the same time, it causes poor ventilation and light transmission in the lower part of cotton plant, which is prone to diseases and insect pests, so that the squares and bolls fall off. When the first fruit branch is squaring, the main stem leaves and all branches of the other nodes are removed except one or two main stem leaves below the fruit branch.

1.2.4 Remove Redundant Bud

The redundant bud that grows in the junction of fruit branch and main stem, consume nutrition, increase square and boll drop off, if possible remove it at any time, remove redundant bud 2 to 3 times in whole growth period commonly.

1.2.5 Removal of Old and Diseased Leaves

If the labor force allows, the yellowed old leaves in the lower part of the cotton plant can be removed in time to increase ventilation and light transmission; the diseased leaves can be removed in time to reduce the disease transmission.

1.2.6 Removal of Empty and Diseased Branches

It is to remove the empty branches and sick branches in time to reduce nutrition consumption and disease transmission.

2　Cotton Chemical Control Technology

Cotton Chemical Control Technology is the applying of growth regulators in the process of cotton cultivation to regulate the vegetative growth and reproductive growth of cotton, to shape the ideal plant type, to establish a reasonable population structure, to improve population ventilation and light penetration, to reduce square and boll shedding, to promote precocity, and to improve yield and quality.

2.1　Commonly Used Growth Regulators

There are many kinds of plant growth regulators used in cotton production, such as "802", mepiquat chloride, chlormequat chloride and Ethephon etc.

(1)"802" is a kind of promoting regulator, which can accelerate the growth and development of cotton, especially increase the vitality of root growth and promote cotton growth.

(2) Mepiquat chloride is a kind of delaying regulator, which can control the growth rate of cotton, promote the steady growth, and make the cotton even fruit setting.

(3) Ethephon is a kind of ripening agent, used in late-maturing cotton field, late boll large proportion cotton field, spraying Ethephon, can promote early maturation of cotton boll, reduce rotten boll.

(4) Flumetralin is a kind of delay regulator, which can achieve the effect of topping by spraying. The use of top-free agent is a more demanding practical technology, in different climate, soil, different varieties, different cultivation methods, different growth conditions, must be tested before application.

2.2 Effect of Chemical Topping Pruning Agents

(1) The use of chemical topping pruning agent can effectively inhibit the growth of cotton plant, control the height of cotton plant and reduce the process of artificial topping pruning.

(2) Inhibition of lateral elongation of fruit branches is beneficial to ventilation and light transmission, increase the efficiency of light energy utilization, and make photosynthetic products more transported to reproductive organs.

(3) It has good control effect on preventing redundant buds sprouting and inhibiting group tips, and can shape good plant structure.

(4) The effect of defoliation agent in cotton field was enhanced, the light in the lower part was improved obviously, the rot boll was reduced, the falling boll was reduced, and the boll opening was more concentrated, and the mechanical harvesting was better realized.

2.3 Key Technologies of Chemical Topping Pruning Agents Application for Cotton

(1) The chemical topping and pruning agent should be applied in combination with conventional control technology. The chemical topping and pruning agent of cotton only inhibited the apical dominance of cotton plant and played a substitute for artificial topping. In cotton, the chemical topping agent and mepiquat chloride can not replace each other. We should pay more attention to the combination of conventional control of mepiquat chloride and increase the boll forming rate at the top of cotton plant and increase the stable yield.

(2) Application time. The first application, according to the cotton

growth, when the height of cotton plant about 55 cm, fruit branches reached 5 sets began to spray. For the second spraying time, the plant height was 75–80 cm, and the number of fruit branches was about 8 sets.

(3) Dosage. Different dosage of products from different manufacturers, the dosage is applied according to the product description, top spray was used in both treatments (mechanical spraying).

(4) In the late period of cotton field, water and fertilizer should be controlled reasonably, so as to avoid the late maturity of cotton. After the second flumetralin spray, irrigation must be controlled for more than 5 days, irrigation should be moderate, avoid excessive water and fertilizer, no drought no irrigation, avoid cotton late maturing.

The apical bud and one leaf at the top of main stem should be removed in topping operation

When the first fruit branch is squaring, the vegetative branch should be removed during pruning operation

Tilapia Natural Reproduction and Farming Technology

1 Preparation of the Breeding Pond

1.1 Selection of Breeding Pond

(1) Location selection: The breeding pond should be selected in a place with good water quality, sufficient water source, convenient irrigation and drainage, and a quiet environment. There should be no tall trees and building houses around the pond.

(2) Area and water depth: The area is generally 300–1,000 square meters. When the broodstock is just placed in the breeding pond, the water depth should be maintained at 1–1.5 meters. When broodstock start spawning, the water depth should be reduced to 0.8–1.0 meter.

(3) Pond shape and soil quality: The shape of the breeding pond is preferably an east-west rectangle. There should be shallow water beaches in the pond to facilitate spawning by broodstock. Aquatic plants should be removed.

1.2 Cleaning and Disinfection of the Ponds

(1) Pond cleaning: Broodstock pond must be cleaned and disinfected before stocking to create excellent breeding environment conditions for broodstock. Generally, the pond water is drained in winter or early spring, the bottom of the pond is thoroughly exposed and cracked, and

then excessive sludge is removed and leveled. At the bottom of the pool, reinforce the pool dike, repair loopholes, and remove weeds.

(2) Detoxification and sterilization: The ponds should be cleared 10–15 days before broodstock stocking. Commonly used drugs are quicklime, bleaching powder, etc. Among them, quicklime is the best, which can kill wild fish, enemy organisms and pathogens in the fish pond, and can also regulate the water quality. Clearing fish pond should be carried out at noon on a sunny day to improve the efficacy. The method of clearing the pond is to drain the pond water, when there is 5–10 cm of water left in the bottom of the pond, use 60–75 kg of quicklime per 667 square meters, first hydrate the quicklime into a slurry, and then splash the whole pond; When the pool water is discharged and 5 –10 centimeters of water are left at the bottom of the pool, use 4–5 kilograms per 667 square meters. After the bleach powder is dissolved in water, the whole pool should be immediately sprayed.

1.3 Injecting New Water and Applying Base Fertilizer

After cleaning the pond, new water should be added to the pond for 1–1.5 meters. During adding water, use a dense mesh filter to strictly prevent wild fish and their eggs and other harmful organisms from entering the fish pond. Then apply base fertilizer to create an excellent water environment and cultivate rich natural bait for broodstock. Base fertilizers include manure (pig manure, cow dung, Sheep dung, etc.), green manure and mixed compost. Generally 500–600 kg of manure per 667 square meters or 400–500 kg of green manure should be applied at one time. Manure must be diluted with water after fermentation and splashed throughout the pool; mixed compost and green manure are stacked in shallow water by the pool to rot and decompose and slowly release the fertilizer effect. After the application of the base fertilizer, when the incubation organisms in the water are multiplying, the dissolved oxygen is rich, and the water quality is good, you can stock broodstock after 10–12 days of clearing pond.

2　Broodstock Stocking

2.1　Broodstock Selection

Broodstock are generally selected twice a year, the first time is to select purebred broodstock, generally based on their traits and body color. The main characteristics of Nile tilapia are: the body color of the fish is yellow-brown, the body has nine vertical black stripes, the ends of the dorsal and caudal fins are black, the caudal fins have 9–10 black vertical stripes, and the ventral and anal fins are gray. The second time is to choose individuals with good body shape, high back and thick body, normal color, clear markings, and good development. Generally, the female weight of Nile tilapia is 250–500 grams.

2.2　Stocking Time

The specific stocking time depends on the local air temperature and water temperature. As long as the water temperature is stable above 18 °C, mature female and male broodstock can be placed in breeding ponds, and when the water temperature rises to 22 °C, natural hybrids can be used to breed fry. In the case of water temperature of 25–30 °C, cross breeding can be performed every 30–50 days. Stocking broodstock should be carried out in sunny and windless weather, and stocking once is better, so that broodstock spawning time can be concentrated, produce same specification fry, which is conducive to fry farming.

2.3 Male and Female Matching Group

The ratio of female to male broodstock is 3 : 1 or 4 : 1.

2.4 Stocking Density

The stocking density of broodstock is 1–2 fish per square meter.

3 Daily Management

After broodstock stocking, the main job is to feed, fertilize and harvest fry. The first is feeding work, feeding artificial feed 1–2 times a day, commonly used feeds are soybean cake, peanut cake, vegetable cake, rice bran, bran, corn flour and so on. It is best to mix several types of feed, and do not feed a single feed for a long time. You can also feed compound feed. The amount of bait is generally 3%–5% of the total weight of the pond fish. After the feed is eaten up quickly, then the feeding amount can be increased appropriately, otherwise the feeding will be reduced or stopped, without waste, to ensure that fish get adequate nutrition. Followed by the work of fertilization and water quality adjustment, the principle of less quantity and more times should be mastered in fertilization. Generally, fertilized manure of 100–200 kg or green manure of 200–300 kg per 667 square meters is applied every 5–6 days. The weather is fine, the water quality is thin, the fish activity is normal, and more fertilization can be applied appropriately, otherwise, less or no fertilization is applied to control the water quality to be of medium fertility. If the water quality is over-fertilized, the fertilization should be stopped, and new water or oxygen should be added immediately to prevent the surfacing of the broodstock and prevent the parentfish spitting out fertilized eggs or fry from its mouth. The third is the job of catching fry.

4　Master Spawning Date of Broodstock

After broodstock stocking, when the water temperature rises above 22 °C, they will begin to spawn and hatch in its mouth. At this time, the pond should be visited frequently to observe the activities of the broodstock and grasp the spawning date and emergence of broodstock. When the water temperature is around 20 °C, the broodstock begin to chase in heat, and soon the females lay eggs, and the males immediately sperm. After the eggs are fertilized, the females immediately suck the eggs into the mouth to hatch. When water temperature is to 25 °C, after 15 days later from laying eggs, fish fry can be emerged swimming around the pond surface. At this time, you can use a soft net to catch the fry.

5　Catching Fry with Soft Fry Net

Fishing fry can generally be carried out in the morning or evening by using small trawls to fish around the pond. The main points of operation are light, slow, and watching carefully, without the need to launch, fishing multiple times, the number of catches is high, the fry is not easy to be injured, and the cross breeding of the broodstock will not be affected by the launch. The fish fry are first put in cages for temporary rearing. After a certain number of fish fry are harvested, they can be put into the cultivation tank for special cultivation. The fry count generally adopts the sampling counting method, that is, select a representative cup to count and then calculate.

6 Fry and Fingerlings Farming Technology

6.1 The Concept of Fry and Fingerlings

Generally speaking, fry refers to larvae of fish with a body length of less than 3 cm; fingerlings refer to juveniles with a body length of more than 3–6 cm, and their ecological habits such as body shape, body color, tissues and organs are similar or similar to those of adult fish.

6.2 Preparation of Fry Ponds

(1) The depth of the pond water should be adjusted with the growth of the fry. The water depth in the early stage is 60–70 meters, the middle stage is about 1.0 meters, and the later stage is 1.0–1.5 meters. The fish ponds should be thoroughly cleaned 12–15 days before the fry is stocked, and the method of clearing ponds are the same as the clearing broodstock breeding ponds do.

(2) Applying basic fertilizer: 10–12 days before the fry enter the pond, the method is the same as that of the broodstock breeding pond.

6.3 Fry Stocking

The water temperature must be stable above 18°C before stocking. Cultivation of 2–3 cm fry, the stocking density should be controlled at 100–120 fish per square meter; cultivation of 4–6 cm fry, controlled at 50–60 fish per square meter; cultivation of fish species over 6 cm, controlled at around 30 fish per square meters. The same fry pond should be stocked with neat and consistent fry; the physique is strong, the swimming is lively, the water resistance is strong, and there is no disease and no injury.

6.4 Farming Management Daily

(1) Feeding method, Take the "three look", "four fixed" feeding principle. The "three-look principle" is: first look at the weather changes, second look at the fish's eating activities, and third look at the water quality. The "four principle" are: fixed time, fixed point, fixed quality, and quantitative. Feeding should be balanced, and the amount of feed should be increased gradually as the fish grows.

(2) Water quality management. With the gradual growth of fry, the depth and breadth of the pond water should be gradually increased. In principle, new water is added once a week, each time it is deepened by 10–15 cm. The 60-mesh sieve is filtered to prevent the invasion of enemy organisms. After 3–4 times of water addition, the depth of the pond water is 1.2–1.5 cm. At this time, the fry are generally about 6 cm. The water quality management index is that the dissolved oxygen content is greater than 3mg / l, the pH value is 7.0–8.2, the transparency is 35 cm, and the water color remains oil-green or dark brown.

(3) Patrol the pond. This is a daily job, one is in the morning, one is in the afternoon and one in the evening. The main task is to observe the activity status of the fry and changes in water quality, check the fish pond for leaks and fish escapes, and remove the fish pond in time Sundries.

6.5　Harvest of Fingerlings

The fry are cultivated in about 30 days, and when the body length grows to more than 6 cm, they can be caught for sale or transferred to commercial fish ponds.

7 Key Points of Adult Tilapia Farming Technology

7.1 Pond Selection

Requires sufficient water source, convenient drainage and irrigation, good light, area of 500–5,000 m^2 water depth of 1.5–2.5 meters, good water retention performance, pond bottom silt is 15–20 cm, with aeration equipment.

7.2 Disinfect the Pond

The water in the pond is 1.0 meter deep, and the whole pond is splashed with 150 kg/mu of quicklime chemical water. On the second day after disinfection, re-inject pond water 1/2 and apply 150 kg /mu of organic manure (livestock and poultry manure) or 250 kg /mu of green manure as base fertilizer to cultivate natural bait organisms.

7.3 Fish Stocking

Tilapia: stocking specifications: 50g / fish; stocking density: 1,000–2,000 fish/mu; crucian carp: 10–20g / fish, 200 fish/mu. Before stocking, the fish should go through a 3%–5% salty water bath for 5–10 minutes.

7.4 Feed Feeding

Under normal circumstances, the daily feed volume is 3% –5% of the total stock of fish, fed twice a day: one is at 9–10 am, account for 40% of the total feeding amount of the day and another one is at 4–5 pm, account for 60% of the total feeding amount of the day. Adjust the daily feeding amount every 10–15 days depending on the fish quantity, to achieve balanced feeding, and the protein of the feed should be maintained between 20%–25%.

7.5 Water Quality Management

Generally, the water injection is changed every 5–10 days, and the pond water is exchanged 15–25 cm each time to maintain a good water quality environment. Transparency is maintained between 25–30 cm. If the weather is clear, there is a slight floating head phenomenon at dawn, indicating that the fish grow vigorously. In high temperature weather and in the middle and late stages of breeding, oxygen should be turned on regularly, and the oxygen should be turned on for 1–2 hours before dawn and at noon. In order to prevent fish diseases and regulate water quality, apply 30 kg of quicklime every 15 days.

7.6 Harvesting

After 4–5 months of rearing, when the tilapia has grown to more than 400 grams and reaches commercial specifications, it can be fished for market.

Key Points of Tomato Cultivation Techniques for High Yield and Good Quality

1 Site Selection and Preparation

1.1 Select Farm Land

For growth tomatoes, the soil should be fertile, rich in organic matter, good air permeability, sandy loam as the best, the terrain should be flat, unobstructed drainage. When choosing a plot, be careful not to continuous cropping with crops such as eggplant, chili, peanuts, soybeans and sesame.

1.2 Prepare the Soil

Prepare the land before planting, by tilling and harrowing it flat, to break up large pieces of earth. At the same time, apply sufficient basal fertilizer, generally apply 30 tons of rotten farmyard manure per hectare, add 750–1,500 kg of superphosphate, combine with ploughing to make fertilizer and soil mix adequately.

Soil preparation

Apply sufficient basal fertilizer

2 Seed Preparation and Treatment

2.1 Select the Seed

First of all, choose varieties with good quality, high yield and disease resistance.

2.2 Sun Seeds

Sun seeds for 5–6 hours one week before sowing.

2.3 Soak the Seeds in Warm Soup

Soak the seeds in clean water for 1–2 hours, then soak them in 55℃ warm water for 15 minutes, at last soak the seeds in clean water for 3–4 hours.

2.4 Accelerate Germination

Place the soaked seeds in an environment of 25–28℃ for 2–3 days. Rinse the seeds with water once a day.

3 Seeding and Seedbed Management

3.1 Sowing

Sowing methods adopts broadcast sowing method. Cover the soil with fine screened soil immediately after sowing. The thickness of the overburden is 0.7–0.9 cm, and the thickness should be consistent. Next, water it well. Then, evenly thin on the bed surface with a thin layer of soil, each square meter with the dosage of 8 grams of 50% carbendazol wettable powder. Finally, cover with a film or sun screen depending on the temperature.

Sow seeds evenly

Cover seedbed with film or net

3.2 Seedbed Management

Seedbed management focuses on controlling temperature and humidity and controlling pests and diseases. The bed soil should be kept moist before seedling emergence, and the mulch should be removed when 70% of seedlings are topsoil, and the soil surface should be kept dry after a good seedling emergence. At seedling stage, control aphid, leaf miner and seedling blight by farm chemical for 1–2 times.

4 Transplanting and Field Management

4.1 Transplanting Density

Row spacing is 60–75 cm, plant spacing is 40–50 cm, 22,500–30,000 plants are planted in hectare.

4.2 Application of Fertilizers

On the basis of applying sufficient basal fertilizer, general every hectare tomato total topdressing quantity is: urea 750 kg, potash fertilizer 1,050 kg.

4.3 Irrigation

Tomato field adopts the pattern of deep furrows and high ridges with good surrounding furrows. Watering should be carried out in the evening, and shallow water irrigation is advisable, to keep the soil moist. If conditions permit, sprinkler or drip irrigation may be used.

4.4 Row Ridging

Combine fertilization and weeding to carry out row ridging for 2–3 times.

4.5 Frame and Tie the Vine

When the tomato grows to 30–40 cm tall, first insert the bracket, then tie the vine. Choose a bamboo pole 1.8–2 meters long and with a thick index finger, and insert it firmly by the side 8–10 cm away from the plant.

Frame and tie the vine

4.6 Pruning and Thinning Fruit

Adopt double stem pruning method, except to retain the trunk, then retain the first leaf axil of the first inflorescence to extract the lateral branches, other branches are removed. Each branch retains 5–7 fruits.

5 Prevention and Control of Diseases and Pests

The focus will be on the prevention and control of diseases and insect pests such as borer (cotton bollworm, tobacco worm, prodenia litura, and beet armyworm), vegetable leaf miner, aphid, whitefly, bacterial wilt, late blight and virus disease. Prevention should be given priority to, and agricultural, physical and chemical prevention and treatment should be carried out in a comprehensive way.

6 Harvest in Time

The ripening process of tomato is divided into four stages: green ripening stage, discoloration stage, ripening stage and complete ripening stage. Export tomatoes can be picked at a discoloration stage (one third of the fruit turns red). Self-feeding and in-place sales can be picked at maturity (the fruits turn red).

Practical Technology on Low Cost Cultivation of Oyster Mushroom

1 Preparation of Before Sowing

1.1 Preparation of Substrate Materials

1.1.1 Cotton Seed Hull

Cotton seed hull it is byproduct of extracting oil factory. It includes the hulls and the short velveteen that is attached to the hull and some broken seeds. Its quality is stable; it is loose and has good permeability; it is easy to get. As a fine material, it has fruiting rate of 120%, sometime reaching 200%.

1.1.2 Rice Straw

Rice straw contains fiber, lignin and waxiness on surface of stalk. Therefore, pre-treatment of straw is the key of successful cultivation. Common pre-treatments are cutting, pulverization, piling for ferment, soaking in lime water, cooking in boil water.

1.1.3 Wheat Straw

Wheat straw is harder than rice straw; waxiness is thicker. It should be exposed to sunlight before cultivation,. Machine or stone grinder is used to pulverize so as to increase porosity and absorption of water.

1.1.4 Other Auxiliary Materials

Common auxiliary materials are lime and gypsum. Note that unsterilized and ferment materials are used to outdoor cultivation; to reduce contamination, usually materials that contain more nitrogen, such as wheat bran.

1.2 Making of Substrate

1.2.1 Formula

① Cotton seed hull 88%, wheat bran 10%, gypsum$CaSO_4$ 2%; ② Rice straw 68%, cotton seed hull 25%, wheat bran 5%, gypsum$CaSO_4$ 2%; ③ Wheat straw 58%, cotton seed hull 30%, wheat bran 10%, gypsum$CaSO_4$ 2%.

1.2.2 Making of Substrate

The weighted cotton seed hull is put on ground to expose to sun; they should be mixed with auxiliary materials and then is sprayed into pre-wetted substrate.

Sun dry material

Formulation of materials

Main material is cotton seed hull. Before ferment, the proportion between substrate and water is 1∶1.5. The ferment compost should be done on concrete ground. The cone holes are made with stick. At last sack and grass curtain are used to cover it for ferment.

Compost substrate　　　　　　　Checking moisture

When compost temperature goes up to 60℃, turn compost. In ferment, plenty of water in substrate gets lost, so proper quantity of water is added according to the situation.

After one week, actinomycetes like snowflake appears in the compost, and the compost become dark brown. Then the compost is spread to disperse waste gas. Water containing is 60%.

1.3　Construction of Cold Frame

1.3.1　Selection of Location

Since cold frame cultivation is done outdoor, temperature and humidity have great influence on it. So, cultivation field is important factor: If select the slope field where soil dries easily, it is not easy to get high yield in cold frame cultivation. The saturated water from soil is more effective than irrigation.

1.3.2　Specification of Cold Frame

(1) Firstly, its length is better to be short, and it usually is five meters. Thus management cultivation is easy to do; fruit bodies grow evenly; it is easy to control pest and disease.

(2) Secondly, cold frame width is better to be narrow. Usually 60 cm

is set for one side picking and 120 m is set for two sides picking.

(3) Thirdly, its depth is subject to local temperature and soil moisture. In wet land, it is 20cm; in dry land it is 40cm.

1.3.3 Digging Cold Frame

One week before cultivation, plot is cleaned and leveled up. Then the cold frame is dug as design. A track or a mound is left between two plots. The bottom of plots is tamped and its walls and its edge are pounded solid to prevent collapsing. Drains are dug around and between the plots. The bottom of drain is slight lower than bottom of plots.

1.3.4 Construction of Shed on Plot

There are various sheds, usually there is arch type. The arch shed offers operation on both sides and it is easy for cultivator to work. It is popular.

After substrate is put on plot and sowing is done, in the place of insufficient shade, shed is built to conduce to mycelia rooting and creeping. Even in woods or orchard, water proof is covered to prevent rain.

2 Sowing

2.1 Preparation of Sowing

2.1.1 Strain and Cultivation Season

In proper season, it is important to arrange time for spawn making.

Polybag spawn production

Checking spawn quality

Recently, cultivation mainly is done in natural climate. The strain should fit for climate and cultivation plan.

2.1.2 Time for Sowing

According to strain type and local climate, time of cultivation is estimated. The four weeks before fruiting is normal sowing time. And then in term of micro climate and cultivation way and weather, sowing time is decided.

2.1.3 Identification of Strain

Quality of strain has a direct influence on yield. Therefore, quality should be checked before sowing. Age of spawn should be reach four weeks, since it is robust and germinates soon. Its apparent features are: mycelia in bottle or bag are white and even in growth and elastic when pressing, with its own smell.

2.2 Sowing in Cold Frame

2.2.1 Procedure

Before sowing, water is sprayed to change moisture according to soil moisture in plot. Lime is sprinkled to disinfect and to dispel earthworm. The ready substrate is put into the plot; and then layer sowing is done. After sowing, surface of substrate is pat flat and then is covered with paper, and shade material or film. At last, shed is covered on the plot.

2.2.2 Sowing Method

Layer sowing: after put a layer of substrate, spawn is sprinkled on it; second layer is put, spawn is sprinkled. Usually 3 layers are put, depending on substrate. The spawn at last layer should be 40% of the total, so that, when mycelia germinate, they can connect together between layers. After sowing, seal the surface as soon to reduce contamination.

Casing soil preparation

To keep humidity is main work after sowing. Remember to ventilate to avoid more moisture. When mycelia grow fully, the film is uncovered and then casing soil is done to keep humidity.

After sowing, shed should be set up as soon for keeping humidity.

3 Management of After Sowing

3.1 Management of Mycelia Growth

3.1.1 Management of First Period

It is the two weeks after sowing. Key work are to control temperature and moisture in substrate so as to avoid high temperature burning mycelia; to prevent from disease and pest; to make mycelia root and grow full of substrate. Therefore, shade material should be thick enough to shield sunlight and should have heat insulation to make dark environment.

In cold frame cultivation, spawn germinates soon; mycelia grow fast; harmful fungi contaminate early. So, the water proof is uncovered for ventilation and substrate is seals tightly to reduce water loss and infect of pest and disease. In the first period water spraying is banned, or else it makes mycelia disappear and makes contamination rise.

3.1.2 Management of Second Period

Mycelia grow full of surface, before the second period management starts. Then work are to boost mycelia to grow into deep part of substrate and to speed up mycelia maturity so ass to turn into nutrition growth.

The detail work is to reduce uncovering times to keep humidity and increase concentration of carbon dioxide that can induce primordium of fruit body forming.

Standard of normal plot in second period is: mycelia grow full of substrate; it sounds bang and is elastic when patting; mycelia are thick on mycelia clod; mycelia skin appears with light yellow drops; the substrate gives off mushroom smell. These features are showing end of mycelia growth phase.

3.2 Management of in Fruiting Period

After mycelia get mature, primordium should be induced to form as soon. Then work is to increase difference of temperature and difference of humidity and light.

Measure to induce primordium: Water proofs at two ends of shed are uncovered at morning and evening to increase ventilation. If plot lose water seriously, spray water to increase. After the uncovering, besides adding light on plot, it can make the mycelia stimulated by difference of temperature between day and night. The measure also makes mycelia grow in alternation of dryness and humidity, so as to hasten mycelia grows to primordium of fruit body. It is noted that after maturity of mycelia, primordium is induced to form by environment factors, such as temperature, humidity, light and oxygen; after one week, primordium can be formed.

After the stimulation, cultivator may uncover the film more times and more time to ensure enough oxygen, In general, ventilation is necessary but prevents too much water loss.

4 Management of in Harvest Period

4.1 Management of before Harvest

Increase oxygen by ventilation; since the second or third day after primordium forming is button stage, it then has weak resistance. Therefore ventilation in the large should not be done. Cultivator uncovers a part of film to ventilate for short time which can ensure air flowing around primordium. Size of intake is subject to degree of its growth. General rule is that primordium should not be exposed too much, or water is lost speedily so as to impede growth.

Control of water and air humidity; Control of water is based on moisture keeping of plot, Water spraying is done in morning and in the evening. Sprayer should be up to make fog drops fall evenly, which can avoid hurt mycelia, or it makes mycelia and fruit body rotten. Water is sprayed to along edges of plot and is led to drain, for keeping 90% air humidity.

By the said works, a part of primordium can grow into maturity. Its features are: pileus edge becomes flat from inner curling; it starts releasing spores. Then it is time to harvest.

4.2 Harvest

Usually, when fruit body is mature, it is time to harvest. The mature standards are edge of pileus become thin from thickness but it doesn't craze. Then the weight of single fruit body is the maximum and it is the most mature. according to situation of market, cultivator harvest in advance or postpone the harvest.

When picking, a hand nip its stipe to turn off and the other hand press its base; if cluster, knife is used to cut off the cluster at the base, but don't pull up the substrate. Cultivator should dig out of the root in plot to avoid inducing pest and disease because of its rotting.

Harvest

Technical Guidance for Whole-Process Mechanization of Maize Production in Ethiopia

Maize is the main food crop in Ethiopia, with a planting area of more than 2 million hectares. To actively promote the whole-process mechanization of maize production is to take the cultivated land, sowing, plant protection, harvest and straw treatment as the key links, promote advanced and applicable agricultural mechanization technology and equipment, cultivate agricultural machinery service market, explore the whole-process mechanization of maize production mode, and realize the increase of agricultural production and farmers' income.

1 Seed

Select high-quality seeds suitable for local planting, with seed clarity ≥98% and germination rate ≥95%. Before sowing, according to the actual occurrence of various local diseases and pests, select the corresponding control agents for seed dressing or coating treatment.

2 Land Preparation

Land preparation include stubble eradication, rotary tillage, harrowing, fertilization and other operations. In order to prepare for sowing, the land should be treated with rotary tiller for stubble eradication after the previous crop is harvested, and plowed, fertilized, rotated and pressed again before sowing to ensure that the surface is smooth, and the soil hardness of the sowing layer is between 300–500 kpa.

3 Sowing

Because the maize growth period is mainly through rainy season in Ethiopia, in order to avoid the long-term soaking of roots by rainwater, ridge planting can be carried out, with the ridge height of 20–30 cm, and seed is planted on the top of the ridge. The planting row spacing should be 60 cm to facilitate mechanized harvesting. When the ground temperature is 8–12℃ and the soil moisture content is about 14%, sowing can be carried out. It is suggested that a multi-functional precision seeder should be used for sowing, so as to realize the one-time operation of spraying, fertilizing, ridging and sowing. The seeder shall meet the following technical requirements: single seed rate $\geqslant 85\%$, cavity rate $<5\%$, seed injury rate $\leqslant 1.5\%$; sowing depth 4–5 cm, plant spacing qualification rate $\geqslant 80\%$; fertilizer shall be applied under or on the side of the seed, with a distance of more than 5 cm from the seed, and the fertilizer strip shall be uniform and continuous; the seedling belt shall have good linearity, and the left and right deviation of the seed shall not be greater than 4 cm.

4 Field Management

4.1 Intertillage Fertilization

In the jointing stage of maize, the high gap intertillage fertilization machine or small field management machinery was used to carry out the mechanized operation of intertillage and topdressing, and the processes of ditching, fertilization, soil cultivation and pressing were completed at one time. The amount of fertilizer applied at each fertilizer outlet of the topdressing machine should be adjusted uniformly. The topdressing machine should have good inter row passing performance. There should be no obvious root injury in the topdressing operation. The rate of seedling injury is less than 3%. The depth of topdressing is 6–10 cm. The topdressing part is 10–20 cm on the row side of the plant. The width of the fertilizer belt is more than 3 cm. There is no obvious broken strip, the soil should be covered tightly after fertilization.

4.2 Plant Protection

According to the occurrence regularity of local maize diseases, pests and weeds, the comprehensive control measures shall be taken based on the requirements of plant protection, Spraying herbicide before seedling should be carried out when the soil humidity is high, evenly spraying to form a layer of drug film on the ground; spraying herbicide after seedling should be carried out in the period of 3–5 leaves of maize and spraying near the ground between rows is required to reduce the drift of drug. In the middle and later period of maize growth, the high gap spraying machinery should be used to carry out mechanized plant protection operation, agricultural aviation operation technology can be promoted to improve the accuracy and utilization rate of spraying chemicals if the conditions permit, and the poisoning of people and animals and pesticide residues of agricultural products should be strictly prevented.

4.3 Drainage

When there is too much water in the field, drainage should be carried out in time with water pump or water conservancy system for automatic drainage.

5 Harvest

5.1 Harvest Time

When the milk-line of maize grain disappears, that is, when the grain is physiologically mature, hard and glossy, it will be harvested. According to plot size, planting mode and operation requirements, the district can process for harvest with appropriate combined harvester and silage harvester

5.2 Technical Requirement

(1) The row spacing of maize harvester should be adapted to the row spacing of maize planting, and the row spacing deviation should not exceed 5 cm.

(2) With mechanized harvesting of maize, plant lodging rate should be less than 5%, otherwise it will affect the operation efficiency and increase the harvest loss.

(3) Maize ears harvest, grain loss rate $\leqslant 2\%$, ear loss rate $\leqslant 3\%$, grain breakage rate $\leqslant 1\%$, ear impurity rate $\leqslant 5\%$, bracts unstripped rate $< 15\%$, stubble height $\leqslant 110$ mm; maize threshing combined harvest, grain moisture content $\leqslant 23\%$.

(4) Corn silage harvesting: shredding qualification rate $\geqslant 95\%$; stubble height $\leqslant 15$ cm; harvesting loss rate $\leqslant 5\%$.

6 Mechanical Supporting Proposal

The equipment is provided in Table below.

Machinery Supporting Proposals for Whole-process Mechanization of Maize Production in Ethiopia

No.	Process route	Appropriate area	Machinery supporting scheme	size (ha)
1	Mechanical land preparation (ploughing, stubble removal, rotary tillage, base fertilizer application, etc.) → mechanical ridge (fertilization, spraying) → mechanical sowing → mechanical field management (plant protection, topdressing, drainage) → mechanical harvesting → mechanical straw recovery.	Small flat area. Such as around Ziway, Debre Zeit, etc.	Equipped with 3 four-wheel drive tractors of 45–55 horsepower, 3 sets of matching ploughshares, 3 rotary tillers, 3 ridges, 3 precision seeder, 2 Mobile sprayers, 2–3 rows of maize combine harvesters, 1 straw bander, and 3 water pumps.	30–60
2	Mechanical land preparation (ploughing, stubble removal, rotary tillage, base fertilizer application, etc.) → mechanical ridge (fertilization, spraying) → mechanical sowing → mechanical field management (plant protection, topdressing, drainage) → mechanical harvesting → mechanical straw recovery.	Large hilly areas, such as from Fiche to Debre Markos, etc.	Equipped with 1 four-wheel tractors of 80 horsepower, 3 sets of 45–55 horsepower tractors, 4 matching ploughshares; 4 rotary tillers, 3 ridges, multifunctional sowing machines or 3 maize precision seeders; 2 Suspension spray boom sprayer, motorized sprayer or electric sprayer; two sets of four rows of maize harvesters or one silage maize harvesters; and 2 straw banders; 4 water pump.	60–100

No.	Process route	Appropriate area	Machinery supporting scheme	size (ha)
3	Mechanical land preparation (ploughing, stubble removal, rotary tillage, base fertilizer application, etc.) → mechanical ridge (fertilization, spraying) → mechanical sowing → mechanical field management (plant protection, topdressing, drainage) → mechanical harvesting → mechanical straw recovery.	Large flat areas, such as Blue Nile River Scoured area in Bahar Dar, etc.	Equipped with 3 four-wheel tractors of 80 HP or above; 3 turn over or mounted plough with the width is 1400 cm to 1800 cm; 3 rotary tiller, 3 multifunctional maize precision seeder; 2 Suspension spray boom sprayer, motorized sprayer or electric sprayer; two sets of four rows of maize harvesters or two sets silage maize harvesters; and 3 straw bander; 4 water pump.	100–300

Practical Agricultural Technologies for Ethiopia (II)

埃塞俄比亚农业实用技术(下)

◎ Written by Wang Li, Chen Xiongzhen, He Wang, Zhou Jilong

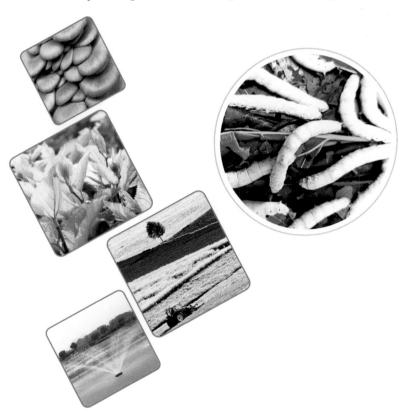

China Agricultural Science and Technology Press

图书在版编目（CIP）数据

埃塞俄比亚农业实用技术.下/王力等著.—北京：中国农业科学技术出版社，2020.12
（非洲农业实用技术丛书；2）
ISBN 978-7-5116-5107-5

Ⅰ.①埃… Ⅱ.①王… Ⅲ.①农业技术—埃塞俄比亚 Ⅳ.①S

中国版本图书馆 CIP 数据核字（2020）第 247443 号

责任编辑　徐定娜　李　雪
责任校对　贾海霞

出 版 者	中国农业科学技术出版社
	北京市中关村南大街 12 号　邮编：100081
电　　话	（010）82105169（编辑室）　（010）82109702（发行部）
	（010）82109709（读者服务部）
传　　真	（010）82109707
网　　址	http://www.castp.cn
发　　行	各地新华书店
印 刷 者	北京建宏印刷有限公司
开　　本	880 mm×1 230 mm　1/32
印　　张	7.5（全三册）
字　　数	345 千字（全三册）
版　　次	2020 年 12 月第 1 版　2020 年 12 月第 1 次印刷
定　　价	88.00 元（全三册）

━━◆ 版权所有・侵权必究 ◆━━

Editorial Board

Editor in chief:

Tong Yu'e

Associate editors in chief:

Luo Ming	Lin Huifang	Hong Zhijie	Xu Ming

Members:

Wang Jing	Wei Liang	Fu Yan	Zhou Min
Yang Yang	Li Jun	Li Jing	

Practical Agricultural Technologies for Ethiopia (II)

Wang Li	Chen Xiongzhen	He Wang	Zhou Jilong

Contents

About Mushroom Cultivation 1

Button Mushroom 13

Oyster Mushroom 17

Cocoon Production and Silk Blanket Process 22

Construction of Three-in-One Fishpond and Its Application

   on Aquaculture 38

Practical Technology Application about Maize Planter 58

About Mushroom Cultivation

Introduction

Mushroom cultivation is a bioconversion process of non edible plant biomass into human food. Agricultural residues for mushroom cultivation are readily available in Ethiopia.

It has been considered as ingredient of gourmet cuisine across the globe; especially for their unique flavor and has been valued by humankind as a culinary wonder. More than 2,000 species of mushrooms exist in nature, but around 25 are widely accepted as food and few are commercially cultivated. Mushrooms are considered as a delicacy with high nutritional and functional value, and they are also accepted as nutraceutical foods; they are of considerable interest because of their organoleptic merit, medicinal properties, and economic significance. However, there is not an easy distinction between edible and medical mushrooms because many of the common edible species have therapeutic properties and several used for medical purposes are also edible.

Mushroom production continuously increases, china being the biggest producer around the world. However, wild mushrooms are becoming more important for their nutritional, sensory, and especially pharmacological characteristics.

Mushroom cultivation in Agarfa was attempted for the first time in 2009 E.C by China's instructor with the collaboration with Agarfa ATVET College. Cultivation and training activities on mushroom cultivation were initiated at the Agarfa ATVET College. Since then, every year experience sharing is being imported to mushroom cultivation. This project established to reach objectives of promoting experience sharing on various aspects of mushroom cultivation and imparting training to instructors, students, farmers and interesting groups.

In recent years, there has been a lot of awareness about mushroom cultivation and its importance. Mushroom growing has created a lot of job opportunities and self employment for both men and women. However, the major constraint in the faster development of mushroom cultivation in Agarfa is in the lack of proper trained manpower. Keeping this objective in view, the present manual has been developed with different activity unit, with all procedures and precaution explained in a simple manner to provide practical work experience to all community as to enable them to undertake mushroom cultivation successfully.

This practical pamphlet is developed to provide you the necessary information regarding the following topics:

(1) Explain the meaning of mushroom.

(2) Recognize commercially cultivated mushroom.

(3) Importance of mushroom.

(4) Optimum growth conditions for mushroom.

(5) Stages of mushroom cultivation.

(6) Materials used for mushroom cultivation.

(7) Procedure.

The Meaning of Mushroom

Mushroom is the fruit body produced by a group of macroscopic fungi (contains no chlorophyll). Majorities of which belong to basidiomycotina of the kingdom of fungi. It is a fungus that can grow in non arable land. It is rich in protein (20%–30%) and a high yielder that remains safe from natural calamities. Furthermore mushrooms are potential contributors to the world food supply since they have the ability to transform nutritionally worthless wastes in to protein rich food. Oyster mushrooms are rather easy to grow on small scale on wide range of substrates and different climatic conditions. Therefore, the transfer of this healthy and environmentally friendly technology has great importance. However, negative attitude and superstitions towards mushroom and lack of scientific research among others has been bottle necks to exploit its potentials. Mushrooms are a source of proteins, vitamins and minerals. They have medicinal properties such as anti-cancer, anti-cholesterol, and anti-tumor functions.

They are useful against diabetes, ulcer and lung diseases. Mushroom is intermediate between that of animal and vegetables in terms of protein content.

On area basis mushroom provides the highest amount of proteins. This is because mushroom could be produced 4–6 times a year. Cultivation of oyster mushroom represents one of the major current economically profitable biotechnological processes for the conversion of plant residues in to the protein rich food, which will help in overcoming protein malnutrition problem in developing countries.

Mushroom is edible mushroom having excellent flavor and taste. It has wide range of substrate growth and temperature adaptation.

Mushrooms are rich in protein compared with other vegetables, and its production can be one of the most promising and highly desirable activities in developing countries to reduce protein malnutrition. Even if, mushrooms has a great value in it nutritional content, its production here in Agarfa is not adopted.

Therefore, this project will undertake to add its share of knowledge on production of oyster mushroom in different locally available organic substrates.

Recognize the Types of Cultivated Mushroom

There are many edible mushrooms growing wild in nature, a few have been domesticated. The five most popular cultivated mushrooms are:

(1) Button mushroom (agaricusbisporus).

(2) Oyster mushroom (pleurotus spp.).

(3) Shiitake mushroom (lentinusedodes).

(4) Paddy straw mushroom (volvariella spp.).

Among these, button and oyster mushrooms are being cultivated in Agarfa and experience sharing here.

Importance of Mushroom

(1) They are good source of high quality protein, vitamins &minerals.

(2) They have medicinal properties.

(3) Mushroom is capable of agro waste degradation.

(4) Mushroom grows independent of sunlight without fertile land.

(5) Mushrooms have a huge export potential.

(6) They offer vast rural employment potential.

(7) Land availability is not much.

(8) Mushroom growing is a women friendly operation.

Design of a Spawn Laboratory

The medium size spawn laboratory should have a total built up area of 20 m × 8 m × 3.6 m. This area will be divided into different work areas like cooking/autoclaving room, sun drying place, inoculation room (3 rooms), store room, store room for strains of spawn and one cold storage room. Cold storage room of 3 m × 3 m × 3.6 m is enough to store the spawn at 4–5 °C. The walls, roof floor as well as door is provided with heavy insulation (7.5–10 cm thickness) and two air conditioner are required to maintain temperature inside the room. Two incubation rooms of 3 m × 6.0 m × 3.6 m with entire surface area (wall, floor, ceiling, doors) insulated with 5–7.5 cm thick insulation. Two air conditioners are required to maintain temperature (25 °C) in the incubation room. The layout of spawn room is followed in table .

The Layout of Spawn Room

Store room			Mixing and filling room	Sterilizing room
Sun drying place for medium bottle piling				
Store room for strains of spawn	Inoculation room	Inoculation room	Inoculation room	Cooling room
				Buffering room
				Inoculation room

Stage of Mushroom Cultivation

There are three stages of mushroom cultivation.

(1) Stock culture.

(2) Pre-culture spawn.

(3) Spawn.

1 Stock Culture

As seed are sown to produce, spores can be used to grown mushrooms. Since this technically difficult, the vegetative mycelium is utilized in modern mushroom cultivation. Spores of mushroom or tissue from the fruit body are capable of growth on suitable organic materials. A combination of organic substance such as starch, sugar, amino acids and inorganic substance fulfill the nutritional requirements of mushroom cultures. Such a combination of nutrients for the growth of microorganism is called a medium. Mushroom cultivation starts by preparing a medium suitable for the growth of the mushroom mycelium.

1.1 Preparation and Sterilization of Media

Depending on nutritional requirement, nutrients are combined and formulated into what is called a medium. The medium is prepared in water and a substance called agar is added to make it solid now an agar medium. Agar is a medium natural substance used to solidify nutrient containing broth.

There are several agar media suitable to grow mushrooms. They contain all the nutrients (sugars, vitamins, salts etc.) required by mushroom.

1.2 Media Components and Equipment

There are many recipes for preparing agar media for mushroom culture, and equipment such as autoclave, incubator, clean beach, tube, flask, needle and loop, etc.

Potato dextrose agar (PDA) is a media commonly used to grow plant pathogen fungi and molds. It is also used to grow mushroom culture. It is prepared in the following ways.

(1) Potato remove the peel 200 g, glucose 20 g, high solidification agar powder 12–15 g, water 1,000 mL.

(2) Potato remove the peel 200 g, bran 50 g, glucose 20 g, high solidification agar powder 12–15 g, water 1,000 mL.

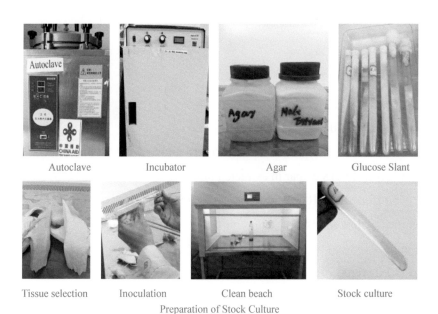

Preparation of Stock Culture

1.3 Procedure

Stock culture of mushrooms can be prepared either by multi-spore or by tissue culture. There are three steps followed. Mushroom growing is a process expansion of home grown spawn. Mushroom mycelium is initially grown on a nutritional agar media. This is then used to make grain spawn. The grain spawn is subsequently used to make the final fruiting substrate. The substrate of stock culture and stock culture are showed in Fig. and stock culture in Fig.

Substrate of stock culture   Stock culture

Step1

(1) Wash 200 g potatoes.

(2) Cut into 0.5 cm cubes.

(3) Boiled in1.2 L distilled water for 15 minutes and filtered.

(4) Add 15–18 g agar in extracted liquid.

(5) Boiled for 10–15 minutes. Add 15 g glucose and 5 g of yeast extract. Allow to dissolve and remove the heat. Pour 50–60 mL into flat bottles.

(6) Covered with cotton wool. Tightened with paper and rubber band.

(7) Sterilized in autoclave at 15–18 kg/cm^2 for 30 minutes.

(8) Slant bottles until agar becomes solid.

Step 2

(1) Take fresh and tender mushroom.

(2) Cut mushroom into half.

(3) Use sterile needle to take small piece of mushroom tissue.

(4) Insert tissue into agar medium bottle and covered with cotton wool and paper.

(5) Incubate impregnated bottle at 20–24℃ with less light for 2–3

weeks.

Step 3

(1) Soak sorghum in water overnight.

(2) Wash sorghum thoroughly with water 2–3 times.

(3) Steam sorghum in pressure cooker for 1 hour and cool.

(4) Add 1% gypsum to prevent sorghum from sticking together.

(5) Fill flat bottles with sorghum up to 3/4 and plug with cotton wool. Sterilized in autoclave for 45 minutes.

2 Pre-culture Spawn

First preparation of medium for pre-culture, select good quality of material stuff, then mix all material stuff completely, sealed with lid after put inside the bottle, finally sterilized by autoclave at 0.2 Mpa for 1.5 to 2 h. Prepare to incubate in shelf room, the procedure is given in Table. The pictures of growing spawn are showed in Fig.

Production of Pre-culture Spawn

Step 1	Select healthy and clean cereal grains
Step 2	Boil grains in water (20–25 min)
Step 3	Remove excess water on sieve
Step 4	Dry grains in shade (2 h)
Step 5	Mix gypsum (1%) on dry wt. basis
Step 6	Fill 250 g grains in glucose/milk bottle
Step 7	Plug and autoclave at 0.2 Mpa for 1.5 to 2 h
Step 8	Inoculate growing mycelium of desired strain using laminar flow
Step 9	Incubate in BOD at $23+20\,°C$ for 20–25 days (shake bottles after 10 days)
Step 10	Pre-culture spawn is ready

Growing pre-culture spawn   Pre-culture spawn

3 Spawn

Polypropylene bags (heat endurance) can be used for spawn. Usually the weight of spawn will be 0.5kg and 1kg, The bags should be of 35 cm× 17.5 cm and 40 cm× 20 cm in size, respectively. Polypropylene bags should have two end sealing. A polypropylene neck (height 2 cm, 4 cm

Mycelia at early stage

in diameter) is placed by plastic ring and folding back the bag after filling the grains. The bags are plugged with cotton ball. These are then sterilized at 0.2 Mpa pressure for 1.5 to 2 hours, then cooling and get to room temperature, prepare to inoculate. The bags are inoculated under aseptic conditions using spawn. Ten to fifteen grams of grains of spawn are used for one bag and one bag of spawn is sufficient for inoculating 25 to 30 spawn bags.

3.1 Production of Spawn

The procedure for production of spawn is given in Table. The growing mycelia at different stages is showed in Fig.

The Procedure for Produetion of Spawn

Step 1	Use polypropylene bags instead of bottle
Step 2	Up to autoclaving (Step 1 to 7) is same as of stock culture
Step 3	Inoculate with 10–15 grams of stock culture per PP bags
Step 4	Incubate at 25℃ in incubation room (Shake bags after 7–8 days)
Step 5	spawn is ready in 2–3 weeks

Mycelia at middle stage

Mycelia at last stage

3.2 Temperature Requirement for Storage and Incubation of Different Mushroom

The basic date is showed in Table.

The Temperatures Required for Storage and Incubation of Different Mushrooms

Items	Lentinula	Pluerotu	Agaricus	Volvariella	Calocybe
Days for colonization of mother spawn	21–23	9–12	21–22	7–8	16–18
Days for colonization in commercial spawn	14–16	9–10	12–15	4–6	11–13
Temperature(°C)needed during colonization	24	24	24	31	24
Temperature (°C) of storage	3–4	3–4	3–4	15	15–16
Growing period of spawn on shelf	60 days	30 days	90 days	<15 days	15 days

3.3 Problems and Solutions

Problems faced during pure culture/ spawn preparation and their solutions are given in Table.

Problem and Solution

Problem	Cause	Solution
Mycelia growth stops at bottom of bag	Heavy grain humidity	Adjust suitable moisture level
Mycelia does not grow completely on the substrate	Contaminated with bacteria and fugi because of improper sterilization	Keep at right time and pressure for autoclave
Mycelia grows very slowly	Lack of enough nutrition, pH, poor quality of gypsum or incubation of improper temperature	Provide balance nutrition for substrate and suitable incubation temperature

Button Mushroom

Button mushrooms are common in meadows, on dung, organic matter rich soil and composted materials in nature. The fruit body of button mushroom can be easily distinguished in three parts: Cap (pileus), stalk (stipe), and root like structures embedded in compost. Growth conditions: Vegetative growth is optimum at 22–25℃ and Fruit takes place at 14–18℃.

1 Material Required

For composting, a variety of agroindustrial residues could be: raw materials. There are variations in the types and quantities of raw materials used for compost making, including the following material stuff, such as cereal straw, cow manure, gypsum, lime and other materials such as chaff cutter, spade, hoe, basket, rubber water tube, measuring tape, wooden boards and pH paper.

2 Compost Preparation for Button Mushroom

Compost is the substrate for mushroom growth obtained after fermentation of agricultural waste (principally cereal straw) by certain beneficial microbes under a particular set of temperature and moisture conditions. The compost provides almost all the essential nutrients, minerals and vitamins required for the growth of mushrooms. Composting is to prepare a uniform suitable for the growth of mushroom but not suitable for the growth of other organism. During composting the microbes break down the insoluble carbon and nitrogen sources present in the straw in to a soluble farm so that they are readily available to the mushroom. Composted organic matter is stable. Moreover, nutrients for competing organisms are decreased.

3　Types of Compost

Compost is of two types. These are synthetic and natural. Cereal straw and chemical fertilizers are used for making synthetic compost. Whereas horse dung and poultry manures are used for making natural compost.

Precaution:

(1) Hygiene conditions should be maintained during entire operation.

(2) Clean the equipment with 2% formalin before use.

(3) Ingredients should be mixed in exact quantities.

(4) Follow the compost turning schedule strictly to ensure that the required end product is achieved at the conclusion of composting.

(5) Excessive wetting of compost should not be done.

(6) Straw heap should not be over compressed while making stack.

(7) Bucket.

4　Procedure

4.1　Preparation of Substrate

(1) Collect straw from nearby farm.

(2) Chop the straw into small pieces (8–10 cm long) with a chaff cutter and transfer in baskets to the composting site.

(3) Select a place with floor as a composting site. The site should have a clean floor with a roof / shed and open sides.

(4) Mix thoroughly the entire quantity of wheat brain, gypsum, and lime in the wetted straw.

Mix the substrates

The substrates completed

(5) Make a heap of about 1m wide and 1m height with the help so spade.

(6) Sowing and mixing the pre-culture spawn in substrate bed, covered with a layer of substrate on the top of bed.

(7) Compressed the heap in to a loose mound by applying slight pressure with the help of wooden boards.

(8) Open the entire heap and remark it a number of times according to the following schedule: Mixing of material and 0 day. Making stack: 1st turning 5th day; 2nd turning 8th day; 3rd turning 10th day. After add gypsum: 4th turning 13th day; 5th turning 16th day; 6th turning 19th day, 7th turning 22nd day; 8th turning 25th day; Filling 28th day.

4.2 Growing Mushroom on Bed

(1) Watering on the mixture after each turning. The turning is carried

out by cutting open with gardening forks, shaking up the manure and then re-stacking in another place on the same site, the different stages are showed below.

| Substrate on the bed | Sewing the seed | Covered with plastic membrane | Appearance of mushroom | Mature of mushroom |

(2) Observe the color of the compost at each turning. The compost is ready for spawning when the color becomes dark brown and there is no smell of ammonia. If a smell of ammonia is present, give 1–2 more turning.

(3) Take a thermometer an insert in the heap at different spots and record the temperatures during composting. Observe the temperature of the stack daily.

(4) Test pH of compost with paper.

(5) Observe the moisture of the compost by compressing the compost between palm and fingers. Water should ooze out of the fingers, but it should not trickle down the fingers (palm test). Moisture of the compost should be around 65%–70%.

(6) Management of Insect-Pests and Diseases

(7) Harvesting, see Fig. below.

The harvesting of button mushroom

Oyster Mushroom

It is wood inhabiting mushroom found in most part of the world. Oyster mushroom can be grown on a variety of plant waste materials containing lignin, cellulose and hemi cellulose such as wheat straw, stalk and leaves of sorghum pearl millet and cotton, maize cobs dried stumps of soft wood trees and vegetable plant residues. They grow on variety of agricultural residue.

1 Material Required

The materials required in cultivation of mushroom, include wheat, wheat bran, sorghum, cotton waste, coffee seed hulls and saw dust.

The materials required

2 Sterilization of Substrate

2.1 Precautions of Substrates Sterilization

Substrate can be sterilized by hot water treatment, steam pasteurization or chemical treatment

Precautions:

(1) Substrate should be fresh and free from moulds.

(2) Chopping of substrate should not be done in the area where substrate is prepared.

(3) Temperature and duration of substrate treatment should be followed strictly.

(4) During bag filling, the temperature of the substrate should not be allowed to excess temperature.

(5) Chop the substrate in to small pieces and soak in fresh water for some hours, depending on the substrate.

(6) Follow any one of the procedures for sterilizing the substrate.

2.2 Procedure of Substrates Sterilization

A. Hot water treatment

(1) Put the chopped substrate in hot water (80℃) for 1–2 hrs.

Dry the substrate

(2) Drain the excess water by putting the substrate on a sieve.

(3) Dry on the sun to reduce moisture.

(4) Squeeze the substrate by hand to check moisture.

(5) Cool the substrate below

28 ℃ and use for filling bags.

B. Steam pasteurization

(1) Wet the straw overnight in a drum.

(2) Fill the pre wetted straw in the wooden boxes/trays and then keep in pasteurization chamber .

Autoclave sterilization

Sterilized substrates

(3) Pasteurize the material with steam at 60–65 ℃ for some hours.

(4) Cool the substrate below 28 ℃ and the use for filling bags.

C. Chemical sterilization

(1) Prepare a solution.

(2) Soak 10 kg fresh chopped straw in the chemical solution for 16–18 h.

(3) Press and cover the soaked straw with a polythene sheet.

(4) Fill the substrate in polythene bags.

3 Inoculation of Spawn

(1) Put 15%–20% of the spawn of substrate weight in the top of substrate bags after sterilization.

(2) Covered the bags with the lid after compressing the spawn.

(3) Transfer it to spawn room.

Inoculation of spawn

Inoculation of spawn

4 Culture of the Mycelium

(1) The room should be kept in a clean, well ventilated situation.

(2) The temperature should be kept around 15℃ .

(3) The humidity should be kept around 50%–60%.

(4) The turning of bags routinely.

(5) Select and discard the polluted bags.

(6) This stage lasts for 30–40 days.

Spawn running of oyster mushroom in bags

5 The Management of Growing Mushroom

(1) The room should be kept in a clean, well ventilated, stable temperature and humidity, and suitable light situation.

(2) Primordial stage of mushroom, this stage needs diffuse sunlight, temperature of 10–15 ℃, last for 3–7 days, open the cover, keep the humidity of 80%–85%.

Mushroom bud stage

(3) Mushroom bud stage, the temperature should be kept around 15–16 ℃, relative humidity of 80%–85%, proper light and ventilation.

(4) Fruit body stage, increase ventilation and sunlight, improve humidity of 85%–90%.

(5) Harvesting, harvest the mushroom when the cap reach 3–5 cm, the times of harvesting may be 3–7 days.

Harvesting

Cocoon Production and Silk Blanket Process

1 Cocoon Production

1.1 Preparation

Prepare rooms, tools and appliances before rearing: rooms, rearing trays, rearing stand, paper or newspaper, plastic film, cleaning net, bucket, basin, chopping knife, chopping board, mountage…

Rearing room, rearing tray, and rearing stand

Rearing tray and rearing stand

Rearing rray

Rearing room

Cleaning net

Plastic film

Black cloth, basin, bucket and newspaper

Chopping knife and chopping board

Mountage for spinning cocoons

Mountage for spinning cocoons

1.2 Disinfection

Disinfection is very important for the successful cocoon production. Procedures of disinfection:

(1) Clean rooms and tools.

(2) Make the 1% bleaching powder solution.

(3) Disinfect with the solution by spaying, immerging and washing.

Cleaning

Making solution

Disinfect by spraying

Disinfect by Immerging

Disinfect by washing

1.3　Incubation and Hatching

Incubation requires 75%–85% humidity and 24–27℃ temperature. The eggs will reach blue stage on about the ninth day of incubation.

Incubation

Eggs before blue stage　　Eggs of reaching blue stage
Egg color before blue stage and reach blue stage

The blue egg stage eggs are kept in dark/black boxes for more uniform hatching on the next day.

1.3.1 The Procedures of Handling Eggs of Blue Stage

(1) Wrap the egg card with paper.

(2) The wrapped eggs are wrapped by black cloth or kept in black box.

(3) The eggs will be exposed to light at 5 a.m. in the morning after 2 nights.

Wrap the egg card

Eggs in black box

Eggs exposed to light

1.3.2 Hatching

(1) The newly developed larva breaks out the egg shell and comes out is called hatching.

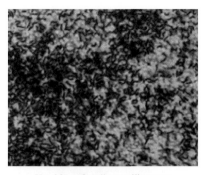
Hatching of mulberry silkworms

Hatching of eri silkworms

(2) Hatching is the beginning of the larva stage.

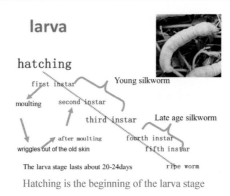

Hatching is the beginning of the larva stage

1.4 Brushing

1.4.1 Identification of Brushing

The process of separating the newly hatched larva from their egg shell and transferring them to the rearing bed is called brushing.

1.4.2 The Procedures of Blushing the Newly Hatched Silkworms

(1) Feed cut leaves to silkworms.

(2) Wait 15 minutes.

(3) Take out the card of silkworm egg and use feather to brush leaves with worms to the worm bed.

(4) Dust fine powder of formalin (2%) to disinfect worm bodies.

(5) Feed them again in 10 minutes after dusting formalin.

Feed cut leaves to silkworms

Wait 15 minutes

| Take out the egg card and brush | Dust formalin fig | Feed again |

1.5 Maintenance of Optimum Conditions for Rearing

1.5.1 The Optimum Temperature and Humidity

It is about 23–28℃ and 70%–85%. The young age silkworms require higher temperature(25–28℃) and higher humidity(80%–85%), and the late age silkworms require lower temperature(23–25℃) and lower humidity(70%–75%).

1.5.2 The Techniques of Adding and Maintaining Humidity Include

(1) Spray or pour water on the floor.

(2) Cover wet cloth.

(3) Cover plastic film.

Cover wet cloth to maintain humidity   Cover plastic film to maintain humidity

1.6 Feeding

1.6.1 Importance of Leaf Quality

The quality of the cocoons harvested depends mainly on the quality of leaves fed during rearing.

1.6.2 Requirement

Young worms require tender leaves while late age worms/require mature leaves; Young worms have to be given chopped leaves. The late age worms may be given entire leaf.

1.6.3 Comsumption of Leaf

The amount of leaves to be supplied for silkworms of 50 laying or 20,000 eggs is about 400 kg for bivoltine race and 350 kg for multivoltine race. Uni and bivoltine require more leaves than multivoltine. During IV and V instars, the silkworms will consume about 85% of the total leaf required .

1.6.4 Frequency of Feeding

Usually feed 3–4 times per day. No feeding during the moulting stage.

Leaf characters for the first instar: tender , yellow-green color ,picked from the top to the third leaf

The young silkworms have to be given chopped leaves

The late age worms may be given entire leaves

1.7 Bed Cleaning

1.7.1 Necessity

It is necessary to remove the litter (which consists of remains of leaves, exuvia of moulted larvae and faecal matter) periodically, and the process of its removal is called bed cleaning.

1.7.2 The Frequency of Cleanings

1st instars': Once during the pre moulting stage.

2nd instars': Twice, once after moult and the second before the next moult.

3rd instars': Three times, first after the moult, second in the middle of the instars' and third just before the next moult.

4th & 5th instars': For shelf rearing, once a day. For floor rearing, once for each instar's.

1.7.3 The Procedures of Cleaning Rearing Bed

(1) Put a net on the rearing bed.

(2) Feed leaves to the silkworms.

(3) Move the net with silkworms to another clean tray before next feeding.

(4) Remove the droppings and the old leaves as rubbish.

Put a net on the rearing bed

Feed leaves to the silkworms

Move the net with worms to another tray

Remove the droppings and old leaves as rubbish

1.8 Spacing

1.8.1 Necessity

Silkworms develop very rapidly from age to age and increase several times their original weight and size in each instar. The total increase in weight from hatching to the end of the 5th instars' will be from 7000–10,000 times. Spacing is very important technique to prevent disease.

1.8.2 Suitable Space

There should be free space of 2 times space occupied by silkworms.

1.8.3 Ways

Spacing is done either independently or along with bed cleaning. Spacing should not be given during the moulting period.

Suitable space

Too crowded

1.9 Care During Moulting

Moulting occurs four times during the larval life. It is a sensitive period lasting for 15–30 hours, during which the worm does not feed but wriggles out of the old skin and comes out with a new and soft skin.

Care during moulting is stopping and resuming feeding at appropriate time that ensures uniformity in growth. And keeping the bed dry and taking anti muscardine measures during moulting will reduce the chance of contraction of diseases during this sensitive period.

(1) Silkworm characteristic before moulting: shrink in size, tight skin, and eat less.

(2) Silkworm characteristic during moulting: no move, no eat, lift head and chest, and △triangle on the head.

(3) Silkworm characteristic after moulting: bigger size of mouth, new skin, and move for food.

(4) Techniques for caring silkworms before moulting: feed a little bit of small size cut leaves, and warm up to 26–28℃.

(5) Techniques for caring silkworms during moulting: no feed, no cover, and dust dry material(lime powder or husk) to silkworm body and bed.

(6) Techniques for caring silkworm after moulting: first feed at suitable time (mostly feed when more than 95% after moulting), second put a net on the bed before feeding and clean the silkworm bed before next feeding.

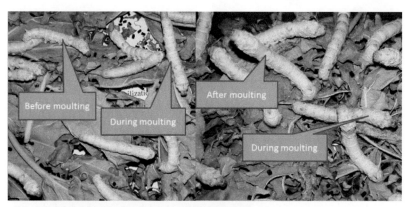

Silkworm characteristic before moulting、during moulting and after moulting

Techniques of during moulting: no feed, no cover, dust lime powder

1.10 Mounting

For providing optimum spinning condition, the ripe worms are transferred to special devices called "Mountage". The process of transferring the ripe worms to the mountages is called mounting. The aim of sericulture is to rear silkworms and provide them with optimum conditions so that they spin a good cocoon with high silk content.

1.10.1 Characteristic Features of a Ripe Worm

The final instars' larva after full growth stops feeding and is ready to spin the cocoon. At this stage it is called the ripe worm. The ripe worm is readily distinguishable by its translucent and yellow color as it does not feed and gut does not have any green color in it. The body shrinks in length.

1.10.2 Process of Spinning

This process takes place about one to two days in multivoltine and two to three days in uni/bivoltines. The fibers from the two silk glands come out through the spinneret.

1.10.3 Methods of Mounting

Hand picking and free mounting methods are widely applied.

Ripe worms: translucent, yellow color and the body shinks in length

Free mounting method   Hand picking mounting method

1.11 Harvesting

Harvesting is a process collection of these matured cocoons from mountages properly. The aim of silkworm rearing is to harvest the cocoons produced and sell them to the reeling agencies.

(1) The recommended time of harvesting: It is the 5th day of spinning for tropical races and 7th or 8th day for temperate races.

(2) Method of harvesting: Cocoons are normally harvested by hands.

Harvested by hands

2 Silk Blanket Process

Silk blanket is very good bedding. It is soft, comfortable and healthy without static electricity. It is easy to process, and the price in China is even better than the silk thread in Ethiopia(the price of per kg silk of blanket in China is 360 rmb which equals 1,440 birr, while the silk thread in Ethiopia is 1,200 birr per kg.). There is a potential big market of silk blanket in Ethiopia. The Chinese in Ethiopia, the rich Ethiopian and the luxury hotels are the potential market for the silk blanket.

The process of making silk blanket is simple and low investment, and it takes less time than spinning silk thread. The procedures of process including the following steps:

(1) Produce cocoon(both mulberry and Eri cocoon is good).

(2) Cut the cocoons to take out the pupae.

(3) Cook shells of the cocoon with 0.5% soda solution for 1 hour.

(4) Wash the cooked shells of cocoon.

(5) Dry the cooked shells of cocoon.

(6) Pull silk by machine or by hand.

(7) Spread silk on the cotton bag on the table.

(8) Use needle and thread to fix the silk to the cotton bag along the 4 sides of the cotton bag.

(9) Wrap the silk into the cotton bag.

(10) Sew to fix the silk to the cotton bag.

Producing cocoons

(11) Cover with quality cotton bag.

(12) Pack.

Cutting cocoons

Cooking

Washing

Drying

Pulling silk by hand

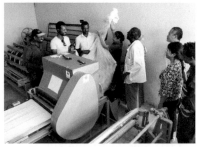
Pull silk by machine

Cocoon Production and Silk Blanket Process

Spread silk on the cotton bag on table

Fix the silk to the cotton bag along sides

Wrap the silk into the cotton bag

Sew to fix the silk to the cotton bag

Cover with quality cotton bag

Pack

Construction of Three-in-One Fishpond and Its Application on Aquaculture

1 Basic Principle

The base to develop aquaculture industry in artificial ponds is the construction of fishponds. In addition to consider sustainable water source, the water retention capacity of soil is one of the most important factor in fishpond construction. Central rift valley of Ethiopia is generally a sandy soil with poor water retention capacity, and the fishponds previously built by several fishery resources and fish research centers in Ethiopia were composed of concrete, with long construction time, high cost and maintain frequently due to crack, and difficult for poor farmers to afford. Therefore the study of our topic is proposed with the idea of constructing fishponds from "triad" local materials and conducting fish farming experiments in the ponds at Algae ATVET College in Ethiopia, which is conducted to investigate feasibility of fishpond constructed from mixture of local materials named "Three-in-One"

and to assess its suitability and capacity for aquaculture. Fishponds are excavated, walls built with three layers: plastic membrane, "Three-in-one" soil and cement pavement. Fish growth in the ponds is evaluated under supplement of different sorts of agricultural residues or agricultural by-products. The "Three-in-One" fishpond technology is found to be low cost, simple to construct, able to retain water effectively and has long service life, and the fishpond also supports fish growth with good farming performance.

1.1 The Meaning of Three-in-One Fishpond

- The cross-section of fishpond' wall has three layers.
- The middle layer of soil mixture also has three type of materials.

What is the meaning of three-in-one fishpond?

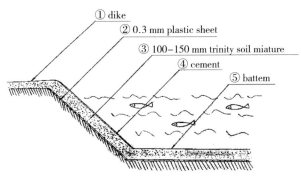

① dike
② 0.3 mm plastic sheet
③ 100–150 mm trinity soil miature
④ cement
⑤ battem

Wall and bottom schematic diagram of "three-in-One" nursery fishpond structure

1.2 The Shape of Three-in-One Fishpond

- The shape in top view is any one if you like.

- The wall is gentle slope which is more than 1:1 in generally speaking.
- The bottom slope is 3‰–5‰ degree from inlet side to outlet.
- In general, the fish pond has four sides with rectangle shape.

1.3 The Materials Needed for Constructing Three-in-One Fishpond

- Common soil
- Termite soil/ quick lime
- Cement
- Teff straw
- Plastic sheet
- Water pipe and valve
- Brick
- Digger: Spade, hoe, or excavator
- Carrier: bucket, wheelbarrow
- Bubble meter

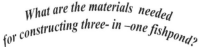

1.4 The Advantages of Three-in-One Fishpond

- Available materials
- Low cost

- Short period for completing construction
- High capacity of holding water
- Can irrigate other vegetation, including vegetable, fruit tree, flower, etc.

2　Procedure of Constructing Three-in-One Fishpond

2.1　Familiar with the Main Body Parts of Fishpond and their Purpose

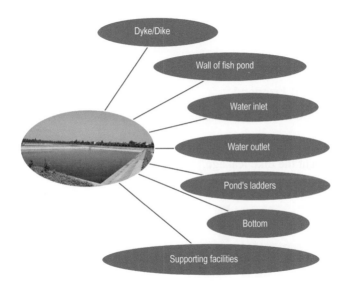

A fishpond with different parts

General speaking, the fishpond has many components with different function as below.

2.1.1　Dyke/Dike

(1) Withstand the greatest water pressure.

(2) Avoid flood.

(3) Used for plantation of green fodder.

Pond with soil dike　　　　　　Pond with concrete dike

Fish ponds with different types of dike

(4) Transportation.

2.1.2　Wall of Fishpond

(1) Enclose waters inside fishpond.

(2) Reduce the water pressure to dike.

(3) Prevent the dike leakage.

The wall of gentle soil slope　　The wall of gentle concrete slope　　Vertical brick wall

Fish ponds with different types of wall

2.1.3　Water Inlet

(1) On the one top side of fishpond for waters entering into pond.

(2) With screen to prevent the entry of wild fish or others wild creature namely filtrate the water.

(3) Prevent the escape of cultured fish.

(4) It is recommended that the pond inlet and outlet be located at opposite ends of the pond to facilitate flushing (good water in and poor water out) when poor water quality becomes an issue.

The water steel pipe with a filter　　Water concrete canal　　　　　　　PVP pipe

Water outlets by using different materials

2.1.4　Water Outlet

(1) Normally locate at the deep end of the pond with the bottom sloping toward it. Most of the ponds used by small scale farmers do not have drains.

(2) Periodic draining and drying of ponds is important because it helps in harvesting fish, eradicating predators, improving the bottom condition of the ponds, and raising production rates.

(3) Pond outlets should have an anti-seep collar and an anchor-collar.

The outlet at the bottom line　　Concrete outlet with wing-wall　　Outlet with filter

Different types of outlet

2.1.5 Pond's Ladders

(1) Usually the ladder locate at the corner area, but sometime at the middle part of long lateral side.

(2) Ladder's number are depend on the size of fish pond, it can be several if the fishpond is so big.

(3) Ladder can let people getting in or out easily.

The ladder at the corner of fish pond   The Ladder near the corner of fishpond

The ladders are located at different place of fish pond

2.1.6 Bottom

(1) Right slope degree of bottom, it can drain up water completely.

(2) Usually there is slope down from inlet side to outlet side.

Waters can flow to outlet by water gravity because above ponds with right bottom slope

2.1.7 Supporting Facilities

Supporting facilities can provide better management for aquaculture, including below item:

(1) **Screen** can prevent wild fish or other predators from getting inside fish pond, as well as from entering grass or other impurities.

(2) **Sieve** is similar with screen, usually it is located in front of outlet gate so that it can prevent fish from escaping from fish pond through the outlet gate.

(3) **Sluice** can control waters entrancing fish pond or not.

(4) **Bird net** can prevent birds from eating fish.

(5) **Water pipe** can deliver waters to ideal place.

(6) **Store and sedimentation pond** can purify water as well as flowing waters to fishpond anytime.

(7) **Store house** can be multi-function building which can store different materials including fish feed, fisheries material, as well as equipment for experiment and office for display management principles, photos, etc.

(8) **Fence** can prevent animals or child from felling into the fishpond, as well as thieves steal fish.

(9) **Overflow gate** can control waters level inside fish pond, therefore fish inside cannot escape from the top of fishpond, as well as the flooding waters cannot erosion the dike of fish pond.

Screen and sluice

Sieve

Net for preventing birds

Water pipe　　　　Store and sedimentation pond　　　Multi-function building

Fence　　　　　　　Overflow gate　　　　　　　Overflow gate

Different supporting facilities for fish pond or aquaculture

2.2　The Theoretical Procedure of Constructing Fishpond

(1) Selecting the suitable site for constructing fish pond.

(2) Cleaning the debris and bush or weeds.

(3) Layout the design of different fishponds and roads.

(4) Digging center part from top to bottom.

(5) Slopping the four sides of fish pond.

(6) Compacting and slopping the bottom area.

(7) Leveling and slopping the dike.

(8) Paving on the pond with suitable plastic sheets.

(9) Digging and constructing the outlet.

(10) Digging and constructing inlet.

(11) Constructing the pond' ladder.

(12) Making the soil mixture and paving on the surface of pond's walls.

(13) Paving cement on soil mixture after making holes.

(14) Level the roads and other place ambient.

2.3 The Practical Procedure of Constructing Three-in-One Fishpond

There is a practical example at Alage ATVET College in Ethiopia which is nursery fishpond with specification 20 m × 10 m×2.5 m, and the wall slope is 1 ∶ 1, and bottom slope is 3‰. The designs for each view are below.

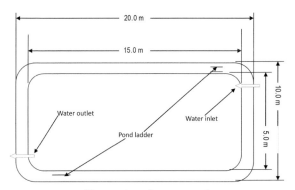

The top view of nursery pond

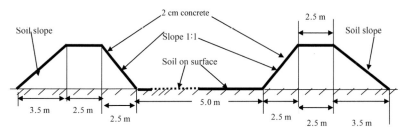

The sectional view of nursery pond

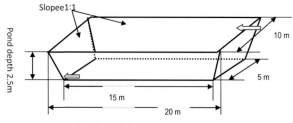

The lateral view of grow-out pond

2.3.1 Selecting Suitable Site

There are some factors related with aquaculture.

(1) *Water supply*

- The water sources must be *reliable and adequate*.
- Good quality water is rich in oxygen, nutrients and free from pollutants.

(2) *Soil type & quality*

The soil type may be sandy soil, clay soil, clayey soils, loamy soils, but the soil should have the below capacity:

- High water retention (holding) capacity.
- Good aeration.
- Adequate nutrient.
- Favorable chemical properties.

(3) *Topography of the site*.

- Gravitational flow of water can be exploited (water can easily enter into the pond).
- Reduce soil excavation and energy consumption.
- Easy to drain water from the pond.

(4) *Other criteria*

- Accessibility.

- Availability of labour.
- Availability and cost of material.
- Availability of marketing outlets and prices.
- Availability of credit and technical assistance.
- Pattern of land and water use.
- Peace and order situation.

Finding suitable environment　　Finding adequate water source　　Checking quality of water and soil

Selecting suitable site for constructing fish pond or fish farm

2.3.2 Cleaning the Site Area

Cleaning the site area, including debris and bush or weeds.

Moving debris and big tree away　　　　Burning the grass and bush

Cleaning the site area

2.3.3 Layouting the Design of Different Fishponds Properly

Layouting the different parts of fish pond

Checking the corner angry of fish pond

Students practice

Layout the design of different fishponds properly

2.3.4 Digging Center Part from Top to Bottom

Digging soil at top

Demonstration for students

Digging soil at bottom area

Excavated soil from top to bottom area

2.3.5 Slopping the Four Sides of Fishpond According the Soil Type

Making a modal of slopping

Method of slopping

Slopping according to requirement

Slopping the wall of fishpond

2.3.6 Compacting and Slopping the Pond's Bottom Properly

Checking the depth of fishpond by using rope

Slopping the bottom area correctly and compacting the margin line

Checking slope degree by using balance

Slopping the bottom area of fishpond

2.3.7 Level and Slopping the Dike

Moving the soil away from dike

Leveling the dike

Checking slope degree by balance

Leveling and slopping the dike

2.3.8 Paving on the Pond with Suitable Plastic Sheets

Measuring the width and length

Cutting the right size of plastic

Paving the fishpond

| Folding the margin of two plastic sheets | Pressuring membrane edge | The model of paved fishpond |

Paving on the pond with suitable plastic sheets

2.3.9 Digging and Constructing outlet

Opening a segment at deepest side   Constructing the outlet   The Model of outlet gate

Digging and constructing outlet

2.3.10 Constructing the Pond's ladder

Opening segment of plastic   Making a steel grid frame   Putting stone under the frame

Construction of Three-in-One Fishpond and Its Application on Aquaculture

Making wood steps   Adding cement mixture   The completed ladder

Constructing the pond's ladder

2.3.11 Digging and Constructing Inlet

Digging a canal for water pipe   Fixing and burying the water pipe   Trying water entrance

Digging and constructing inlet

2.3.12 Making the Soil Mixture and Paving on the Surface of Pond's Walls

Making soil mixture   Paving on the wall of fishpond   Making holes on muddy surface

Making the soil mixture and paving on the surface of pond's walls

53

2.3.13 Paving Cement on the Surface of Soil Mixture

Cementing the pond' wall　　　　　　The fishpond cemented

Paving the fishpond by using cement

2.3.14 Leveling the Roads and Other Place Ambient

 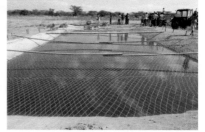

Leveling the dike and sowing grass　　　The landscape of fishpond

Leveling the roads and other place ambient

3　Aquaculture

3.1 Fertilizing and Enriching the Waters Inside Fishpond

Adding compost   Adding green grass togethe   Filling inside with waters

Fertilizing and monitoring waters

3.2 Transporting Fish Seeds by Using Close System

Filling fish inside with oxygen   Transporting fish by car with air- conditioner   Arriving at fishpond site

Transporting fish seeds by close system

3.3 Releasing Fish and Culturing it

Adjusting water temperature inside bags   Releasing fish   Feeding fish

Releasing and feeding fish

3.4 Harvesting Fish

Harvesting fish

4 Key Points

(1) Ponds that are poorly designed or constructed may cause:

- Ponds could not hold water.
- Ponds could not be drained completely (leading to incomplete harvests and poor production on subsequent production cycles).
- Ponds could not be filled or drained by gravity, and dikes could collapse.

On the other hand, well-designed and constructed ponds are easily managed and maintained, leading to less "down time" due to failures and more efficient operation and production.

(2) Efficient organization of support facilities in relation to the pond system is paramount (critical) significance in the overall developmental planning and operation of the farm.

(3) Pond must be checked and maintained every year.

- Repair the damage place, such as filling the holes and crack area.
- Moving the grass if it grow on surface of wall.
- Moving out the muddy on surface of bottom area after 2–3 years.

(4) The bottom area can not be touched by sharp materials.

(5) The fish density should be properly, otherwise fish will be dead

because high density of fish may cause low concentration of dissolve oxygen inside waters if there is no aerator.

(6) In order to reduce production cost of aquaculture, agricultural residues or agricultural by-products such as soybean meal, oil seed cake, wheat or rice bran, maize flour, bone and blood meal, fish offal meal, and poultry dung can be mixed and used as fish feed. The ingredients were abundant in the area and relatively of cheap prices and efficient to promote fish growth.

5 Expected Results

(1) The construction technology of "Three-in-one" fishpond has been expanded in three departments at Alage ATVET College, and we recommend this technology could be used elsewhere in Ethiopia and even in other parts of Africa where fish culture in pond is a challenge because of water retention problem.

(2) The coefficient of feed will be low and fish yield is also high under good routine management if the quality of waters and feed are good.

Practical Technology Application about Maize Planter

1　General Description

This type of maize planter is designed and made in Agarfa ATVET College, which is very suitable for village farmers to use easily and efficiently. It can plant every seed (grains such as corn, wheat, soybean and peanut) at the same depth with the same quantities by this manner. Therefore I especially introduce it to Ethiopian people and it have the Following advantages.

(1) The maize planter can be made easily from local available wooden timber and other materials.

(2) It can control the number of seeds to be planted every site: either one, two or three as you want.

(3) It can control the soil depth of the different seeds needed to be planted.

(4) It is not only suitable for maize, also suitable for peanut, soybean.

(5) It can work without the need of electric power and other source of energy to start up the planting process, and easy to carry out by every local farmer.

Therefore, the present maize planter will save money, time and human labor.

2 The Principle

This maize planter comprises a planter tool with a trapezoid container (2) for receiving seeds, and a seed control gate (3) which enable to store seeds thus ready to enter in the seed controller (4). A seed controller structure is mounted under the container, it consists of two parallel slide boards (5) and (6) with controlling hole (7) and (8) which make backwards and forwards action, can control the number of seeds, the seeds go down into the opener (9) to be planted in the soil. The drive mechanism with spring (10) and drive stick (11) which drive the seed controller through rope (12) and tackle (13); and drive the opener through a level (14) and wire (15), and the depth controller (16) is attached under the drive stick. the handles (17) on the top of the container to be used to operate the seed planter.

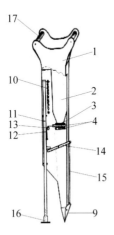

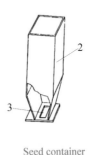

Seed container

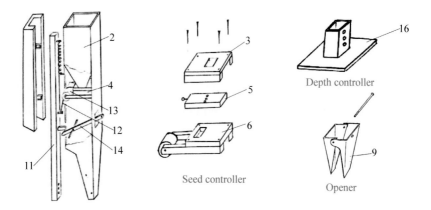

Seed controller

Depth controller

Opener

3 Designing and Making of Machine

3.1 Designing of the Machine

This maize planter consists of four sections:

(1) Seed container.

(2) Seed controller.

(3) Drive system.

(4) Opener.

Components and Functions of Maize Planter

Components	Main functions
Seed container	To contain seeds
Seed controller	To control number of the seeds to be planted
Drive system	To drive the seed controller and opener mechanism
Opener	To open the soil for the seed to be planted

The mechanical structure of the planter is showed by the following drawing.

Practical Technology Application about Maize Planter

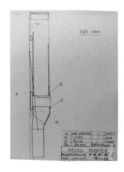
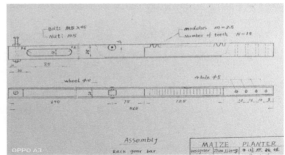
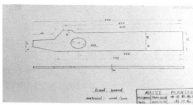
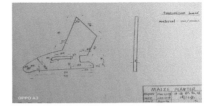

3.2 Making Process

(1) Materials preparation.

Making this maize planter need some materials as follow:

Timber, three-ply board, square steel tube, steel plate Spring, nail, and so on.

(2) Tools preparation.

Wood working machine, saw, hammer, screwdriver, electric drill, and so on.

(3) Making process.

See the picture as below.

4 Operation and Adjustment

4.1 Adjustment of Number of Seeds

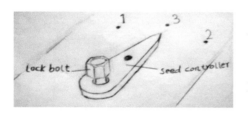

Before planting, you have to adjust the number of the seeds, one, two or three as you want.

First loose and turn out the lock bolt smaller, then turn the seed controller to the mark 1, 2 or 3, the mark 1, 2 or 3 means one, two or three seeds will be planted. After adjustment, tie the lock bolt again.

4.2 The Plantation Depth Adjustment

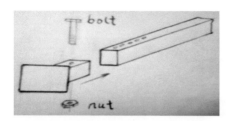

Different seed needs different soil depths for planting. so have to adjust the planter depth controller before planting.

Loose the nut, remove the bolt out. There are more holes on the push rod, put on the bolt in the different holes, will get different depths of the planting. Choose suitable position, then put in the bolt and tie it properly.

4.3 Operation

The operation is simple, just catch hold of the handle and push down the planter, the seeds will go down into the soil through opener.

5 Preservation and Storage

If the planter is to be out of service for a long period of time, it is necessary to preserve and storage.

(1) Cleaning all the surface parts, in order to prevent any corrosion and erosion.

(2) Do not use water to clean, may damage the planter.

(3) Put it on dry and cool place.

Practical Agricultural Technology Book Series for Africa

Edited by Center of International cooperation Service, MARA

Practical Agricultural Technologies for Zimbabwe

津巴布韦农业实用技术

◎ Written by Zhang Shihong, Wu Chunhua, Luo Dengfeng, et al.

China Agricultural Science and Technology Press

图书在版编目（CIP）数据

津巴布韦农业实用技术 / 张世洪等著. —北京：中国农业科学技术出版社，2020.12
（非洲农业实用技术丛书；3）
ISBN 978-7-5116-5107-5

Ⅰ.①津… Ⅱ.①张… Ⅲ.①农业技术—津巴布韦 Ⅳ.①S

中国版本图书馆 CIP 数据核字（2020）第 247444 号

责任编辑　徐定娜　李　雪
责任校对　贾海霞

出 版 者	中国农业科学技术出版社
	北京市中关村南大街 12 号　邮编：100081
电　　话	（010）82105169（编辑室）　（010）82109702（发行部）
	（010）82109709（读者服务部）
传　　真	（010）82109707
网　　址	http://www.castp.cn
发　　行	各地新华书店
印 刷 者	北京建宏印刷有限公司
开　　本	880 mm × 1 230 mm　1/32
印　　张	7.5（全三册）
字　　数	345 千字（全三册）
版　　次	2020 年 12 月第 1 版　2020 年 12 月第 1 次印刷
定　　价	88.00 元（全三册）

━━◆ 版权所有·侵权必究 ◆━━

Editorial Board

Editor in chief:

Tong Yu'e

Associate editors in chief:

| Luo Ming | Lin Huifang | Hong Zhijie | Xu Ming |

Members:

| Wang Jing | Wei Liang | Fu Yan | Zhou Min |
| Yang Yang | Li Jun | Li Jing | |

Practical Agricultural Technologies for Zimbabwe

Zhang Shihong	Wu Chunhua	Luo Dengfeng	Li Xiaoqiang
Li Dongsheng	Wu Weijun	Duo Likun	NuerShafa
Yan ShouGen	Su Zhengming	Liu Weibo	

Contents

Zimbabwean Crop Farming Practical Training Manual 1

Horticulture Training Manual 10

Key Techniques of Free Range Chicken Raising
　Management 17

The Way of Using of Double-barreled Mobile Milking
　Machine 27

The Technique of Pig Breeding and Chinese Sausage Processing ... 34

The Technique of Rabbit Breeding 42

Practical Technology for Raising Tilapia in Cages 50

The Prevention and Control Knowledge of Zoonotic Diseases
　Preventive Methods in Zimbabwe 56

Prevention and Treatment of Disease Animals 61

Handbook on Agricultural Machinery Maintenance
　Technicals 67

Mechanization Technology of Saving Cost and Increasing
　Efficiency in Wheat Production 76

Rotor Tiller 82

Zimbabwean Crop Farming Practical Training Manual

1 Maize

English Name: Maize
Scientific name: *Zea mays* L.
Shona: Chibage/Magwere
Ndebele: Umumbu

1.1 Background

Maize is a strategic crop in Zimbabwe because it is the staple food (Sadza) crop and it doubles as a cash crop. About 64% of maize is for human consumption, 22% for livestock and 14% for industrial use. The crop accounts for 14% of agricultural Gross Domestic Product, and remains essentially an industrial input, and the primary source of food and nutrition security. National maize production peaked at about 2.8 million metric tonnes in 1985 with average crop yields peaking at about 2.2 tonnes/ha that same year. Maize yields in the smallholder sector peaked at 1.4 tonnes/ha. In 2010, national maize production was about 1.3 million tonnes with an average yield of 0.7 tonnes/ha. Zimbabwe's annual commercial maize requirements are approximately 2.2 million tons. It has been a government policy over the past years to promote maize production as an effort towards food self-sufficiency. However, during the last 10 years, Zimbabwe has been importing maize and has been facing serious shortages of this staple crop.

1.2 Growing Conditions

1.2.1 Soils

Sandy clay loams and heavier soils are most suitable.

1.2.2 pH

The crop requires a optimum soil pH range of 5.5 to 6.5.

1.2.3 Temperature

The optimum temperature range for maize growth and development is about 18–32℃. Maize is intolerant to cold; it must be grown in the hot summer months of September to April.

1.2.4 Rainfall Requirements

Annual total rainfall of 500–750 mm or more is required for maize. If severe moisture deficit coincides with tasseling and silking, crop failure is to be expected.

1.3 Varieties Selection

NRI and II: late and medium maturing varieties.

NRIII-IV: early maturing varieties.

For the best variety for your area, consult your local or nearest extension office.

Seed rates: General rate is 25 kg/ha, but with small seed more area can be covered.

1.4 Planting Dates

Dry planting: 2–3 weeks prior to beginning of rains (From mid-October).

Rain planting: planting with the first effective rains (From mid-October).

Green mealies: beginning of August.

1.5 Plant Population

Plant 2 seeds per station.

45,000 plants/ha (90 cm×25 cm) for high rainfall areas of NR I and NR II.

37,000 plants/ha (90 cm×30 cm) for low rainfall areas of NR III and IV.

22,000 plant/ha (1.5 m×30 cm) on tied ridges for dry areas of NRV.

Varieties with short stature can be planted at higher plant populations.

1.6 General Fertiliser Recommendations

Fertilizer	Natural region II	Natural region III	Natural region IV
Compound D/ Maize	300 to 400 kg/ha	250 to 300 kg/ha	200 to 250 kg/ha
Ammonium Nitrate	300 to 350 kg/ha	200 to 300 kg/ha	150 to 200 kg/ha

1.7 Weed Management

Maize should be maintained weed-free for at least 2–6 weeks after establishment. Mechanical control: use of ploughs, cultivators, tractors and hand hoeing. Use of herbicides to kill weeds. Cultural weed control: involves use of practices such as early planting, intercropping and crop rotations.

1.8 Pest Management

Pest	Damage	Control
Maize stalk borer	Horizontal rows of holes on the leaves emerging from the funnel of the plant and frass (droppings).	Deep ploughing and destruction of stover. Use thiodin or dipterex.
Cutworms	Plants are cut soon after germination at ground level.	Use baits of maize meal and insecticide, and/or of pyrethroid sprays applied in bands over the rows, Karate.
African Armyworm	Eat voraciously often leaving only stems and mid-ribs of leaves.	Contact insecticides such as malathion, trichlorfon, endosulfan.
Termites	Base of the stem and buttress roots are attacked, resulting in lodging.	Dress seed with Regent 500 FS. Use Regent 200 SC in a 150 cm band in-furrow.

1.9 Disease Management

Common maize diseases in Zimbabwe are grey leaf spot (GLS), maize streak virus (MSV), leaf rust, *Turcicum* leaf blight (TLB) and ear rots. Generally these diseases can be controlled through the use of resistant varieties, rotating cropping system and treat seeds with fungicides for protection against soil-borne fungal diseases.

1.10 Harvesting

Harvesting will be done when the maize reaches physiological maturity. Maize takes 200–220 days from planting to dry naturally to 12.5% moisture content. It can, however, be reaped at 25% M.C. or less and dried. Maize can either be dried artificially or naturally. In natural drying, the cob is left on the plant while the sheath can be opened to speed up drying.

Field visit on Chinese maize trial plot by Mr Nyamangara, Zim-China CATAP Project Coordinator General, Director of Department of Education & Farmers Training of MoA

Official inspection on Chinese maize trial plot by Mrs Tong Yu-e, Director General of CICOS, Ministry of Agriculture and Rural Affairs,PRC, together with her team

Chinese maize varieties public introdution to visitors on Zimbabwean Agriculture Show 2019 at Harare Agricultural Show Ground

Maize field interview at DR&SS, Harare for ZBC Zim-China Mega Project

Maize trials plots harvesting at Gwebi by Chinese expert with Zimbabwean colleagues

Good harvest for farmers in Muronbedzi assisted by Chinese expert

Maize production training

Maize variety trial at Gwebi

Reported on ZBC Television

2 Wheat

English Name: Wheat.

Scientific Name: *Triticum aestivum* L.

Shona Name: Gorosi.

Ndebele Name: Ingqoloyi.

2.1 Background

Wheat is the second most important agricultural commodity in Zimbabwe for food security in terms of quantity and calories consumed (per capita wheat consumption requirement is 28 kg per annum). According to the Grain Millers Association of Zimbabwe (GMAZ), 2019 monthly wheat requirement stands at 38,000 metric tons translating to an annual requirement of about 456,000 metric tons. With an annual production of 160,663 metric tons, it means Zimbabwe is less than 40% self-sufficient.

2.2 Growing Conditions

2.2.1 Soils

Fertile, well drained sandy and clay loams.

2.2.2 pH

The crop requires an optimum soil pH range of 5 to 6.

2.2.3 Temperature

The optimum temperature range for wheat growth and development is about 16–23 ℃.

2.2.4 Rainfall Requirements

Annual total rainfall of 500–600mm or more is required.

2.3 Varieties Selection

Ncema, Dande, Insiza, Nduna, Smart, Sahai, Kame, Kana etc. are nice varieties.

Seed rates: Drilling: 100–120 kg/ha, Broad casting: 120–160 kg/ha.

2.4 Planting Dates

Highveld: 15th–25th of May, depending on frost prevalence in the area.
Middleveld: 1st to 15th of May.
Lowveld: 1st of May.
Plant spacing: 20–25 cm between rows.
Planting depth.
Heavy soils: 4–5 cm; Sandy soils: 5–6 cm.

2.5 Rolling

4 weeks after planting and germinated, rolling once by tractor.

2.6 General Fertiliser Recommendations

Basal: Compound D @ 400 kg/ha.
Top: Ammonium nitrate @ 400 kg/ha.
However, farmers are advised to sendtheir soils for analysis so as to get the exact fertilizer requirements for their particular soils.

2.7 Weed Management

It is very important during the first 4–6 weeks after crop emergence.
Mechanical and chemical methods are commonly used.
Post-emergence Herbicides: Fulisade W/ Fulisade Super, MCPA, Basagran, Puma Super.

2.8 Pest Management

Quelea bird is the major pest of wheat in Zimbabwe. Its control is done by the National Parks, who should be informed by the farmers. Other pests include wheat stem sawflies, hesian flies, aphids and cutworms. Dimethoate 40 EC is used to control aphids. 85% Carbaryl is used to control army worm, caterpillars.

2.9 Disease Management

Rusts and Smuts are the major diseases of wheat.

Cultural control methods such a s field hygiene, crop rotation, regulating sowing dates are recommended. Use of fungicides like Shavit will reduce the spread of the diseases.

2.10 Harvesting

Should start when the back of the head has turned from green to yellow and the bracts are turning brown, about 30 to 45 days after bloom, and seed moisture is about 30% or less.

Manual harvesting is practiced by cutting the crop with a sickle or knife.

A combine harvester can also be used.

Training and planting wheat

Micro-jet irrigation

Rolling for more tillerings

Site inspection by delegations led by Mr Guo Shaochun, the Chinese Embassador

Field visit by Mr. Koza, the Deputy Director for Mechanization, MoA

Inspection by Mrs Mazvita, the Head of Agriculture Research Institute, DR&SS

Wheat Field Day orgnized by Chinese Agricultural Expert Team

Chinese technology introduced on Wheat Field Day

Gwebi 2019 Chinese Wheat Field Day reported on *Herald* News, the mainstream media

Horticulture Training Manual

1 Introduction

Grafting is a method of propagation, in which, the plants are obtained from a vegetative portion of the mother plant instead of seeds. It involves no change in genetic makeup of the new plant. All the characteristics of the parent plant are reproduced in the daughter plant due to exact duplication of chromosomes during cell division. Thus, the plants are true-to-type in growth, ripening, yield and quality.

1.1 Budding

Budding is a method in which only one bud is inserted in the rootstock.

1.1.1 T-budding

A horizontal cut about 1/3rd the distance around the stock is given on the stock 15–20 cm above the ground level . Another vertical cut 2–3 cm in length is made down from the middle of the horizontal cut and flaps of the bark are loosened with ivory end of the budding knife to receive the bud. After the 'T' has been made in the stock the bud is removed from the bud stick. To remove the shield of bark containing the bud, a slicing cut is started at a point on the bud stick about 1.25 cm below the bud, continuing slicing to about 2.5 cm above the bud.

A second horizontal cut is then made 1.25 to 2 cm above the bud, thus permitting the removal of the shield piece. The shield is removed along with a very thin slice of wood. The shield is then pushed under the two raised flaps of bark until its upper horizontal cut matches the same cut on the stock. The shield should fix properly in place, well covered by the two flaps of bark, but the bud itself exposed. The bud union should be wrapped with polythene strip to hold the two components firmly together until the union is completed. After the bud has healed, the top of the stock is cut off just above the bud. Thus, forcing the new bud to grow. The polythene strip will disintegrate in several weeks. If it does not, or you use a non-biodegradable tie, you will need to cut the polythene band. T budding can be performed at any time of the year provided cell sap flows freely.

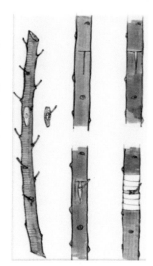

Budding

1.1.2 Patch Budding

A rectangular or square patch or piece of bark about 1.0–1.5 cm broad and 2.5 cm long is removed from the rootstock at about 15 to 20 cm from ground level. A similar patch with a bud on it is removed from the bud stick taking care not to split the bark beneath the bud. This patch is then transferred to rootstock and fixed smoothly at its new position and tied immediately with polythene strip.

1.1.3 Chip Budding

In this method, one about 2.5 cm long slanting cut is given into the

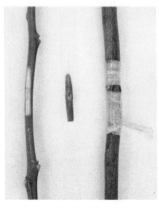

Chip budding

stock followed by another cut at lower end of this first cut, in such a way that a chip of bark is removed from the stock. The bud from the scion wood is removed in the same way so that it matches the cuts given in the rootstock. This chip with a bud on it is fitted smoothly into the cut made in the rootstock taking care that the cambium layers of the stock and scion unite at least on one side. The bud is then tied and wrapped with polythene strip, to prevent drying up of the bud.

1.2 Grafting

Grafting is another method of vegetative propagation, where two plant parts are joined together in such a manner that they unite and continue their growth as one plant. The upper part of the graft is called the scion and the lower part is called the stock. In this method, the scion twig has more than two buds on it.

1.2.1 Whip Grafting

First, a long, smooth, slanting cut of about 4 to 5 cm long is made on the rootstock. Similar cut is made in the scion wood exactly matching the cut given in the rootstock. The scion having 2 to 3 buds is then tightly fitted with the rootstock taking care that the cambium layer of at least one side of the stock and scion unites together. This is then wrapped with polythene strip.

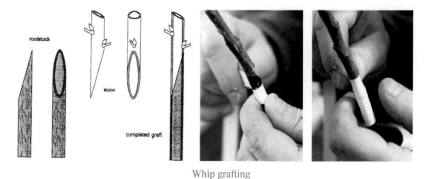

Whip grafting

1.2.2 Tongue Grafting

This is simply a whip graft with the addition of a tongue. The tongue helps to hold the graft in position, making it easier to tie. In addition, the scion can be left sitting in position to be tied by the second member of a grafting team . There is increased cambial contact between the rootstock and scion, and the graft is mechanically stronger than that of whip grafting.

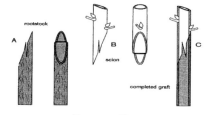

Tongue grafting

For this graft, make two cuts in the stock as shown in. First, make a long, sloping cut through the rootstock shoot, as for a splice graft. Then, place the knife edge straight across the surface of the first cut, about a third of the way down from the top. Cut into the face of the first cut on a slightly steeper angle, more or less straight down the stock. Make the depth of the tongue about one-third of the length of the first cut. Similar cut are made in the scion wood exactly matching the cut given in the rootstock, aligning the cambial areas on at least one side. Wrap and cover.

1.2.3 Cleft Grafting

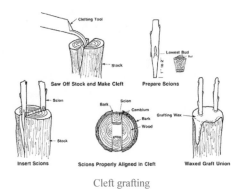

Cleft grafting

The stock up to 8 cm in thickness can be grafted with this method. The rootstock to be grafted is cut squarely and smoothly with secateurs or saw. It is then split in the middle down to about 4 cm. The bud stick having 3 to 4 buds is trimmed like a wedge at the lower end with outer side slightly broader than the inner side. The lower bud on the scion should be located just well in to the stock making sure that the cambium layers of both the stock and scion are perfectly matched. All exposed surfaces are waxed or coated immediately. Even though the stock exerts sufficient pressure to hold the scions, wrapping the stock is still needed to ensure a tighter connection and less chance for the scion to be bumped out of the stock. Cleft grafting is done during dormant period.

1.2.4 Side Grafting

Side grafting

A three-sided cut about 4×1.25(base) cm is made on the rootstock at a height of about 15–20 cm from the ground level and the bark of the demarcated portion is lifted away from the rootstock. A matching cut is also made on the base of the scion to expose cambium. The scion should be prepared well before the actual grafting is done.

The healthy scion shoots from the last mature flush are selected for this purpose. The selected scion shoots should have plump terminal buds. After the selection of the scion shoots, remove the leaf blades, leaving petioles intact. In about 7 to 10 days the petioles shall drop and terminal buds become swollen. At this stage the scion stick should be detached from the mother tree and grafted on the stock. The prepared scion is inserted under the bark flap of the rootstock so that the exposed cambia of the two components are in close contact with each other. The bark flap of the rootstock is resorted in its position. The graft union is then tied firmly with polythene strip.

After the completion of the grafting operation, a part of the top of rootstock is removed to encourage growth of the scion. When the scion has sprouted and its leaves turned green, the root stock portion above the graft union should be cut away.

1.2.5 Veneer Grafting

In this method, a shallow downward cut of about 4 cm long is given on the rootstock at a height of about 15–20 cm from the ground level. At the base of this cut, a second short downward and inward cut is made to join the first cut, so as to remove a piece of wood and bark. The scion is prepared exactly as in side grafting. The cuts on the rootstock and scion shoot should be of the same length and width so that the cambial layers of both components match each other.

Veneer grafting

Then, the prepared scion is inserted into the rootstock and tied security with polythene strip. After the union is complete the stock is cut back.

1.2.6 Saddle Grafting

Both rootstock and scion should be the same diameter. For best results, use saddle grafting on dormant stock in mid-to late winter. Stock should not be more than 1 inch in diameter.

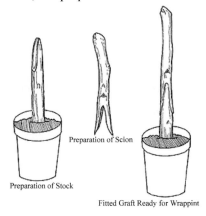

Saddle grafting

1.2.7 Bark Grafting

The rootstock is severed with a sharp saw, leaving a clean cut as with cleft grafting and a vertical slit is made. The bark is loosened, the scion is set, and then the bark is pressed in place and the trunk is tightly wrapped. Alternatively, nails can be inserted. All exposed surfaces should be waxed.

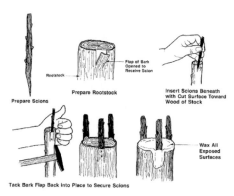

Bark grafting

Key Techniques of Free Range Chicken Raising Management

1　Breeding and Management of Young Chicks

To produce meat as the main purpose of chicken, this period is 1 to 28 days old; to lay eggs as the main purpose of chicken, this period is 1 to 42 days old.

1.1　Feeding Density

The density of brooding on ground is 25 chicks/m^2; It is 30 chicks on the net; it is 50 chicks in the cage.

1.2　Water

After the chicks out of the shell, after 12–24 hours of observation in the storage room, put into the nursery, and then observe 1–2 hours to start drinking water, namely after the chicken fleece dry 3 hours to 24 hours between the arrangement of the first drink. Use a small vacuum and warm water to drink for the chicks. The water temperature should be the same as room temperature. Let the chicks drink 0.05% potassium permanganate water or 5% glucose normal saline or 7%–8% sucrose water, clean drinking water or clean drinking water with added vitamins and minerals. For chickens transported long distances, oral rehydration salts should be added to drinking water to help regulate fluid balance.

After the first drink, should not cut off water.

1.3　Feed and Feeding

Chicks should be fed completely equivalent compound feed. Quality raw materials such as high quality soybean oil, expanded corn, expanded soybean and imported fish meal should be added into the chick mouth feed.

Generally in the first three hours after drinking, when the chicken showed a strong appetite, let the chicken start to eat. Spread the special chick feed on cardboard or plastic sheet, and let the chicks eat freely to ensure that each chick is half full. After feeding, let the chicks drink freely. After 3 days of age, feed with material barrels and feeding trough containers, 6 times a day before 2 weeks of age, including 1–2 times at night; Feed 5 times a day when you are 3–4 weeks old and 4 times a day after 5 weeks old. Let the chickens feed for 45 minutes at a time, but let each chick feed at the same time. For chicks after 4–5 days of age, add gravel once a week, and mix 450 grams of gravel into the feed for every 100 chickens at a time, increasing by 20% per week.

1.4　Temperature

Suitable breeding temperature is the temperature that the chicks feel comfortable, and chicks related to the age. When the room temperature of the brood house is lower than the appropriate temperature of the chicks, it will be heated up. Instead, cool it down. By heating, the house temperature was maintained at 32–35 ℃ at 1–3 days of age. Every 7 days, decrease by 2–4 ℃. On the 28th day, the temperature in the house dropped to about 18–21 ℃ , and the temperature was maintained thereafter. After 5 weeks of age, the heating stops when the natural temperature reaches 18℃. Raise young in hot summer, when the temperature exceeds the appropriate temperature, by opening ventilation and water curtain device for cooling.

1.5 Humidity

The appropriate humidity (relative humidity) is 50%–70%. When exceeding the appropriate humidity range, the humidity should be reduced through ventilation and timely cleaning of feces. Instead, increase humidity by sprinkling water, etc.

1.6 Illumination

Illumination affects chicks in two ways: one is the duration of illumination, and the other is the intensity of illumination.

In the first week, the illumination was 24 hours per day. After the second week, the illumination was 12–18 hours per day. The night was low light illumination, with the illumination intensity of $3-5 W/m^2$.

1.7 Ventilation

To keep the air fresh through natural ventilation and forced ventilation, 0–2 weeks old, ventilated several times a day, 10–30 minutes each time, in order to discharge waste gas, adjust the humidity inside the house. After 3 weeks of age, strengthen ventilation to reduce the ammonia concentration in the shed to below 25 mg/L.

2 Feeding Management in the Early Stage of Stocking

To produce meat as the main purpose of chicken, this period is 1 to 28 days old; to lay eggs as the main purpose of chicken, this period is 1 to 42 days old.

2.1 Change the Chicken House (Transfer Chicken Group)

Chickens are transferred from the breeding house to the medium chicken house. Adjust the two houses to the same temperature before changing houses. Stop feeding 6 hours before the flock is transferred, and drinking and feeding again after entering the new house. Before and 3 days after chickens are transferred, add 1–2 times the normal amount of vitamins and electrolyte solution to the drinking water. On the day of the shift, light is provided for 24 hours to ensure that food and water intake is not affected. Sick, weak, underweight and stunted chickens can be selected and eliminated at the same time.

2.2 Male and Female Feed Separately

Males and females feed in two groups. Separate male and female at the same time as changing the chicken house.

2.3 Feeding Density

In this period, it is advisable to raise chickens in the ground or net flat. Chickens can be raised 10 per square meter on the ground and 20 per square meter on the net.

2.4 Water

Let the chicken drink freely. According to the needs of chicken growth and metabolism, add glucose, electrolytes and vitamins.

2.5 Feed and Feeding

Replace the feed at the breeding stage with the feed at the pre-stocking stage. For example, among the 1,000 kg of feed, there are 645 kg of corn, 180 kg of bean cake, 80 kg of fish meal, 80 kg of silkworm chrysalis, 10 kg of bone meal, 1.3 kg of eggshell meal and 3.7 kg of salt. It contains 192 kg crude protein and has a metabolic energy of 125.6 MJ.

It takes a week to change the feed. Replace the chicken feed with medium chicken feed at a rate of 15%–20% per day until all the chicken feed is replaced with medium chicken feed. Gravel was added to the feed at a rate of 5 kilograms per 1,000 chickens per week.

2.6 Conditioning

Conditioning refers to the process of giving specific instructions or signals in a specific environment to gradually form conditioned responses or generate habitual behaviors. The training of free range chickens includes feeding, drinking water, far grazing, homing, perch, caging, egg laying and emergency hedging. Unify signal as far as possible, reduce signal type.

2.7 Adaptation

After a period of adaptive training, the chickens gradually adapt to the external climate and environment, and develop the habit of Grazing and homing, then enter the stocking period and graze all day.

3 Feeding Management During Stocking Period

The main purpose of the chickens was to produce meat, and the stocking period was from 57 days old to market. The main purpose of the chicken is to lay eggs, and the stocking period was from 71 days old to post-production.

3.1 Stocking Scale and Density

The stocking scale should be around 1,500 animals per group. The stocking density should be controlled within 3,000 animals per hectare.

3.2 Water

Let the chicken drink freely. Set up a drinking point for every 200 chickens, or place an automatic drinking fountain. Pay attention to the cleaning and disinfection of drinking water.

3.3 Feed

Free-range chicken feed was prepared for artificial supplementary feeding of free-range chicken. The proportion of raw materials used in the preparation of feed for free-range chickens for meat use is roughly as follows: grain feed accounts for 50%–70%, bran feed for 4%–5%, plant protein feed for 15%–25%, animal protein feed for 2%–7%, mineral feed for 1%–2%, additive for 1%–4%, and oil for 1%–4%.

For the direction of egg is free range chicken feed preparation, requirements for the approximate proportion of various feed raw materials are: grain feed 45%–70%, bran feed 5%–15%, plant protein feed 15%–25%, animal protein feed 5%–10%, minerals 5%–7%, dry grass powder 2%–5%, trace element additive 1%.

3.4 Supplementary Feeding

According to the wild feed resources to determine the amount of supplementary feeding. General supplementary feeding can account for

2%–5% of body weight, if laying hens, the supplementary feeding per day is 40–80 grams. In addition, 25%–30% green feed can be added according to the total amount.

The type of supplementary feeding should be determined according to the type of wild feed, if it is mainly weeds, supplementary compound feed, if it is mostly starch feed, pay attention to supplementary protein feed.

If wild feed is abundant, supplementary feeding time is generally scheduled for the evening. In the case of insufficient wild feed, 35% of the supplementary feed on the day of supplementary feeding in the morning and 65% of the supplementary feed in the evening were provided.

When grazing is not possible due to weather conditions, the feeding ration shall be 100%.

It is prohibited to feed moldy, spoiled or contaminated feed.

3.5 Grazing Time

In winter (dry season) grazing occurs after sunrise and in summer (rainy season) before sunrise. The homing time should be no later than before sunset, and should be as early as possible. Generally, the stocking time should not exceed 8 hours.

3.6 Rotation Grazing

The grazing land is divided into several grazing areas. Different batches of free-range chickens should be grazed in different areas. At each site, after a batch of chickens has been raised, the pens and surrounding areas should be disinfected and left empty and cleaned for at least 4 weeks before they can be used again.

3.7 Guard against Animal Damage

Build a isolation net to prevent animal damage. If the grazing land is large, then nylon net can be used to build a mobile isolation net according to the community; Put rat trap and raise goose to prevent rat damage; Set up wind wheel straw air defense eagle; Adopt the double hunting method and the hunting dog method to prevent weasel harm.

4 Comprehensive Disease Prevention and Control

4.1 Biosecurity Precautions

All biosecurity measures can be taken to prevent entry of the outlanders and vehicles, to prevent entry of exotic birds and their products, to prohibit livestock mixing, to prevent wild birds and wild animals from entering poultry houses, and to eliminate pests and rodents.

4.2 Sanitary Disinfection

Daily strengthening of the disinfection of feeding utensils, after each batch of chickens are put out, the poultry house shall be thoroughly disinfected after cleaning and cleaning. Use natural exposure and other possible means to disinfect the grazing grounds.

4.3 Immune

According to the local epidemic situation of chicken diseases, appropriate vaccines, immunization procedures and methods are selected for the inoculation of chicken diseases. The general immunization procedures are as follows:

(1) 1-day-old chickens were injected with marek's disease liquid nitrogen vaccine.

(2) Inoculate 7-day-old chickens with both Newcastle disease and infectious bronchitis combined with polyvalent freeze-dried vaccine and oil vaccine.

(3) Immunization against bursa bursa vaccine in 14-day-old chickens.

(4) 20-day-old chickens, Newcastle disease and infectious bronchitis combined vaccine.

(5) Immunization against bursa bursa vaccine at 25 days of age.

(6) Immunization against chicken pox and avian influenza vaccine at 35 days of age.

(7) Immunity to Newcastle Disease I strain of 55–60 days old chickens.

4.4 Parasite Control

After 20 to 30 days of initial grazing, half ascaris tablets were used in each chicken, and then once every 20 to 30 days, one ascaris tablet was used in each chicken. Grind the tablets into powder and mix them with the feed.

4.5 Waste Disposal

After the free range chickens are come to market, the bedding and excrement should be removed once, and the suitable site should be selected. The high temperature composting treatment should be covered with thick plastic film.

For the packaging of veterinary drugs and other items as well as discarded veterinary drugs and other wastes, it is necessary to select an appropriate place for centralized destruction and disposal. Do not abandon at will, pollute the environment.

4.6 Epidemic Management

In the event of an outbreak, it shall promptly report to the local veterinary administrative department, follow the arrangement, and take quarantine, culling, destruction, disinfection, emergency immunization vaccination, blockade and other restrictive measures in accordance with the law.

The Way of Using of Double-barreled Mobile Milking Machine

1 Milking Operation Method

(1) Turn on the power of milking machine; turn off the vacuum switch of vacuum tank, and start the vacuum pump, then the finger of vacuum gauge will rise stably.

(2) Adjust the vacuum pump to make the finger point to normal vacuum degree position(47kPa), turn on the switch of vacuum pump, then you can start to milk.

(3) The adjustment method of vacuum pump.

- Disassemble the cover of vacuum stable machine, unscrew the nut of head of the vacuum stable pressure machine, and press the valve core of vacuum stable machine several times to know if the valve core is locked by dust or other dunghill. (Normally, the valve core can turn up and down).

- Circumgyrate the cover of valve core to adjust the vacuum degree. When circumgyrating, you can observe the change of finger of vacuum apparatus. When the finger of vacuum apparatus points the normal position, screw the nut quickly. Then observe the finger of vacuum apparatus, after adjusting several times, you can make the vacuum degree correct.

Note: The vacuum should be started normally, and the performance sounds is also normal.

If the vacuum degree is too high, the teat of cow will be damaged. If the vacuum degree is too low. it will influence the quantity of milking directly.

2 The Order of Milking

2.1 The Preparation of Milking

(1) Clean the breast and teat of cow, and then make it dry. You had better clean the teat with the water with biocide.

(2) Before cover the milk cup, you should milk by hand 2–3 times. Then put the teat in the special cup to observe if the teat is normal.

(3) Clean and make the breast and teat of cow dry. After confirm the teat is normal, you should cover the claw in 1 minute.

2.2 The Method of Covering Milk Cup

(1) First hold the claw horizontal and make the 4 pieces of milking lining liner fall naturally. Then the milking lining liner can close the 4 mouths of-claws with the gravitation of stainless steel cup. Hold the claw horizontal, put the claw under the breast of cow and open the switch of claw by the forefinger of the hand which should always hold the claw horizontal. (If the claw is declining, it will make the milking lining liner in the declining side air escape to influence the normal performance of milking machine.)

(2) Hold the head of milk cup with the thumb, third finger and last finger of the other hand, and leave forefinger and middle finger to guide the teat put into the milking lining liner. Make the stainless steel cup upright. Before doing it, you should disassemble the rubber pipe of

bottom of milking lining liner to make sure the stainless steel cup will not air escape when put it upright, then cover the milking lining liner on the teat quickly. (If the cow has one teat, which cannot milk, you should use the special rubber teat cap in the mouth of milk cup to avoid air escape.)

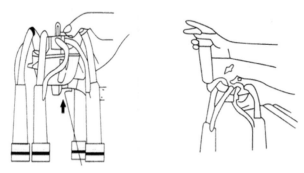

The operation of covering milking cup

3 Milking Process

The time of milking of each cow is different and usually. It costs 3–5 minutes to milk. When see the four teats are almost finished milking, hold the claw with.one hand. And pull the claw ahead, so that it can make the milk left in the area of back breast clean. Pull the claw ahead again, if you find there is no milk left. You can take the cup off.

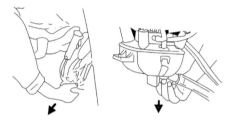

The operation of taking the cup off

4 The Method of Taking the Cup off

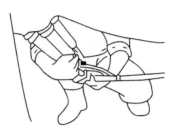

The way of pulling out four milking cup

(1) When the four teats are finished milking, you should take the cup off at once. When taking it off. You should pull the switch of claw down, and close the vacuum.

(2) One hand to close the vacuum switch of claw, and the other hand to hold the four milk cups pull out declining.

Note: please don't milk after finishing milking, or the breast of cow will have sickness.

5 The Correct Method of Installation Rubber Milking Lining Liner

Put the milking lining liner in stainless steel milk cup. Please pay attention to the arrows in the head of milking lining liner and the part of connecting the pipe, and they should be in same line. When pulling the

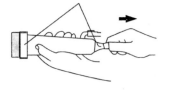

The installation rubber of milking lining liner

milking lining liner in the stainless steel cup, please don't distort is, but pull it straightly. After installation, please examine the two arrows are in the same line.

6 Cleaning Method

After finishing the milking, you must start to clean, after finishing the milking, if you don't clean the milking equipment at once. The bacilli will propagate quickly.

6.1 The Using of Cleaning Agent

The component of the milk is complicated. Therefore, you should use the special cleaning agent to clean the milking machine. And it also will not damage the rubber part of milking machine.

The function of different cleaning agent

Type of cleaning agent	Main function	The effective temperature of cleaning agent	The thickness of cleaning agent
Alkalescence cleaning agent	Milk fatness Milk protein velum	Above 40°C	0.1%
Acidity cleaning agent	Mineral substance	Above 40°C	0.1%

6.2 The Cleaning Order (Daily Cleaning)

(1) Clean the claw, milk pipe, and milk barrel by water.

(2) Cleaning preparation: use a big pot with full water, put and claw in the pot, turn on the vacuum pump. And absorb the water into m Ⅲ barrel from claw. Turn off the vacuum switch of vacuum tank, open the cover of milk barrel, then pour the water.

(3) Circulation cleaning of cleaning agent: Pour the cleaning agent with correct thickness to about 70°C water and homogenize, put the claw in the pot, open the vacuum switch to absorb the hot cleaning agent into milk barrel. After absorption, pour the cleaning agent back to the pot, then circulate the absorption until the temperature fall to 40°C.

(4) Clean with normal temperature water 2–3 times, absorb the water from claw into milk barrel, and clean the cleaning agent in the equipment.

(5) Pour the water, make it dry, and put the milking equipment in clean and airlessness place for the next use.

(6) Normally. The alkalescency and acidity cleaning agent is used circular. 3 days alkalescency, Then 1 day acidity cleaning agent.

7 Familiar Fault and the Dealing Method

7.1 Normal Fault and Dealing Method of Seeing Table Following

Familiar fault and the dealing method

Fault phenomenon	Reason	Eliminate method
The vacuum pump cannot be started	• The electromotor cannot work. • The vacuum pump has fault.	• Test if the electromotor has the pressure, if the connect line of motor is cut (or not replace), and if the pressure of power are suitable for the pressure of motor. • Notify our company, and we will send the technician to deal.
The fault of claw	• Incorrect placing position of gasket ring. • Damage of gasket ring.	• Install the gasket ring correctly. • Change the gasket ring.
Vacuum degree is too low	• The pump oil seal is damaged. • Muffler is blocked. • Tie-in of vacuum pipe is air escape.	• Change the pump seal. • Disassemble the muffler, clean with water or hammer the rust. • Screw the tie-in of parts, and change the damage rubber ring.
The finger of vacuum apparatus swings not stable	• The vacuum stable pressure machine is not clean and blocked. • Vacuum apparatus is damaged.	• Remove the dirty thing of stable pressure machine and clean it. • Change the vacuum apparatus.

7.2 The Pressure Issue

If the pressure is not stable, see following as:

The pressure fault and dealing methods

Fault	Dealing
The switch of the vacuum is not turned off.	Turn off the switch of vacuum
The tie-in of pipe is air escape	Screw the tie-in
The rubber parts are damaged and air escape	Change the rubber parts
The pressure is not stable	Make the pressure stable
The rotate speed of motor is not quick enough	
The stable pressure machine is unscrew	Adjust the stable pressure machine again

7.3 The Motor Issue

If the motor cannot be started, see following table :

The motor cannot work and deal methods

Fault	Dealing
The motor cannot work	Examine if the motor has pressure, if the connect line is cut(or not replace), if the pressure of power are suitable for pressure of motor, if the motor is damaged, or we should change the motor.

7.4 The Motor and Pump Issue

If motor and pump cannot work, see following table :

The motor cannot work and deal methods

Fault	Dealing
belt of motor is damaged	change the belt
motor has fault	notify our company to change

33

The Technique of Pig Breeding and Chinese Sausage Processing

1 Pig Breeds and Breeding

A profitable pig production enterprise should be based on good animals of any improved breeds. Such as Large white. Yorkshire. Landrace. Duroc. Hampshire. Berkshire. Indigenous breed of Mukota. There are two breeding methods which are Pure-breeding and Cross breeding. Breeding methods recommended as following:

Different pig breeding production models

Sire(♂)	Dam(♀)	Offspring	Terminated Sire(♂)	Utilization
1	Pure breeding			Pure breed
Large White	Large white	Large white		Pure breed
Landrace	Landrace	Landrace		Pure breed
Yorkshire	Yorkshire	Yorkshire		Pure breed
Duroc	Duroc	Duroc		Pure breed
Hampshire	Hampshire	Hampshire		Pure breed
Mukota	Mukota	Mukota		Pure breed
2	Crossing breeding			
Landrace Large White	Yorkshire	LY(♀)	Duroc(♂)	DLY piglet
Yorkshire	Landrace Large White	YL(♀)	Duroc(♂)	DLY piglet
Hampshire	Mukota	HM(♀)	Duroc(♂)	DHM piglet
Berkshire	Mukota	BM(♀)	Duroc(♂)	BHM piglet

2　Management of Piglets

2.1　Navel Cord Care

The area where the umbilical cord is broken can be a passage way for disease-causing organisms into the body of the newborn. Treat the navel cord by wetting with iodine solution. You may apply using cotton wool. Restrain the piglet by grasping it over the shoulder or back. Cut off the navel cord leaving about 1 inch.

2.2　Iron Supplementation

(1) Piglets are born with low reserves of iron in the body, and sow milk is low in iron. Iron is required for normal blood formation and the transportation of oxygen. Piglets need iron supplementation to prevent anaemia (Piglet anaemia). Iron can be supplied from several sources: ① Place clods of red soil in the pen. Take care to get soil from an area that is not contaminated with worms; ② Buy iron tablets and give to the piglets. Caution: piglets can cough them out; ③ Buy an iron solution and rub it on the sow's teats. Use iron injections. An experienced person should administer the.

(2) Procedure for injection. Read the label and instruction for the iron product you are using. Select a clean syringe and needle and fill the syringe. Grasp and hold the pig by one of its rear legs. Clean the injection site with a cotton swab containing a disinfectant. Push the needle with a little jab through skin at the cleaned site. Inject the proper dosage slowly into the muscle. If you are to inject into the neck muscle. ① Put the piglet between your knees; ② Stretch its head to one side; ③ Inject the iron into the muscle on the side of the piglet just off the top line.

2.3 Clipping of Needle Teeth

Restrain the piglet by grasping the head with one hand. Force the mouth open using fingers on the same hand near the back edges of the mouth. Be careful that you do not choke the piglet. Use sharp pliers taking care not to injure the gums. Hold the clippers as perpendicular as possible to the teeth. Completely cut off the teeth as close to the gum as possible. After clipping the teeth on one side turn the pig to give access to the teeth on the other side of the head. Disinfect the pliers after working with each litter of piglets.

2.4 Tail Docking

Hold the piglet suspended by the rear legs with one hand. Using a sharp sterile knife cut off the tail to leave 1/2 inch from the place where the tail joins the body. Disinfect the wound. Disinfect the knife after working with each litter of piglets.

2.5 Identification

Piglets can be identified; Notching should be done as follows.

Hold the piglet by the head and use a sharp knife or razor blade to remove a V shaped amount of tissue from the edge of the ear. Remember to notch the correct position of the ear. The method may cause some bleeding. Treat the wound with iodine or some other antiseptic. Disinfect the knife or use a new razor blade after working with each litter.

2.6 Castration

Hold the piglet by the hind legs as shown. Clean the scrotal area with a detergent or any disinfectant. Grasp the testicles and push them upwards to tighten the skin make a cut down the length of each testicle. Cut only through the skin and white membrane. The cut must be at the lower end of the scrotum to allow easy drainage of the blood. Pull the testicle through the opening, twist and scrape the cord. Treat the wound with an antiseptic e.g. iodine to prevent infection. Recording of Events between Farrowing and Weaning this is a very important management operation. After farrowing record the number of piglets born alive and those born dead. Record the number of female and male piglets. After identification weigh the piglets and record the weight of the litter at birth. At 3 weeks of age weigh the piglets again, Weight at three at birth. At 3 weeks of age weigh the piglets again. Weight at three weeks gives an indication of the milk producing ability of the sow and her mothering ability.

2.7 Fostering

If there is a sow which has farrowed within 3 days adjust litter size for the number of functioning teats or milking ability of the sow and move the larger piglets to the foster sow. Move piglets before they are three days old but make sure they have received colostrum from their mother before you transfer them. To ensure that the foster sow does not recognize and reject the fostered piglets cover the smell of the piglets so that it (the sow) cannot recognize its young by smell. You may mask the smell by: ① Smearing all the piglets including her own piglets with a strong-smelling substance like garlic; ② Soaking all the piglets thoroughly in a salt solution. Observe the foster sow as you go through this process to ensure that it is not battering the fostered piglets.

2.8 Creep Feeding

The main purpose of creep feeding is to allow the piglets to get used to the solid diet they will consume after weaning. If the piglets are used to the solid diet, they will be less likely to experience digestive problems at weaning. Creep feed also helps to augment the nutrients that the piglets will be getting from the dam's milk. The milk that the piglets get from the dam is normally not enough to support high growth rate especially after the third week of lactation.

There is very little incentive for the piglets to consume the creep feed during the first three weeks of lactation because milk production will be increasing hence their nutritional requirements will be met by the milk. After the third week milk production will no longer be increasing hence there is an incentive for the piglets to consume the creep feed.

3 The Processing Technology for Making-family Chinese Sausages

(1) The processing technology for making pig casings.

① Selecting Pig intestine which come from health slaughtered pig. The Pig intestine is fresh color and no smell; ② Separation of fat meat from the intestine; ③ Throwing inside parts of intestine away, cleaning intestine outside and inside; ④ Soaking the intestine in the warm water for 1–2 day; ⑤ Wipe the intestines, making intestines clean with warm water; ⑥ Curing with salt for preserving casings.

(2) The processing sausages.

Cutting: Cut the lean and fat meat into small size of 0.5cm slice.

Washing: put some salt to lean meat, and washing the lean meat with cleaning water for separating the blood from lean meat.

Mixing: mixing the lean meat. Fat meat salt and all of stuff together for 2 hours.

Formula1: Lean meat 8 kg, fat meat 2 kg, sugar 500 g, salt 300 g, monosodium 30 g, liquor100 g, ginger powder 10 g, garlic powder 150 g.

Formula2: Lean meat 3.5 kg, fat meat 1.5 kg, sugar 250 g, salt 150 g, monosodium 15 g, liquor50 g, ginger powder 10 g, garlic powder 10.

Formula3: lean meat 4 kg, fat meat 1 kg, sugar 250 g, salt 150 g, monosodium 15 g, liquor 50 g, ginger powder 10 g, garlic powder 10, Chinese white pepper powder (胡椒粉)30 g, chili powde100 g, Chinese red pepper,(花粉)powder100 g.

Cleaning: Putting the casings in warm water and cleaning.

Producing: Put the meat inside of casings.

Drying: putting the sausages under sunlight about one day, then putting the sausages in shadow place which wind can blow the sausages for more drying for about 15 days.

Collecting and preserving: After 15 days, collecting the sausages hang on the wall in your house. You can preserve the sausages more than half year.

Cooking: Put some sausages in your container with water, let the water boil, after 10 minutes ,take out the sausages from a container, cool it for 10 minutes, cut it into small piece, finally you can taste it's delicious.

Chinese sausages making-family processing procedure

Procedure	Illustration
1 Make casings 1.1 Selecting and cleaning intestine 1.2 Separating the fat meat from the intestine	
1.3 The fresh casings for Preserving	
2 Select lean and fat meat, Cut small size of 0.5 cm	
3 Mixture with all of raw materials according to formula	
4 Putting the meat into the casings you can use the mouth of water plastic bottle	
5 Tightening and dry	
6 Cooking and tastes	

Illustration of pig pens

Structure illustration	Illustration
Dry Sow pens	
Lactating Sow pens	
Lactating Sow pens	
Weanner pens	
Grower and Finishing Pens	

The Technique of Rabbit Breeding

1 Selecting Breeds for Breeding

After choosing the appropriate breed to rear, it is important to select the animals that will be used in the breeding programme, based on specific selection criteria. The breeds of rabbit include: California. New Zealand. Chinchilla and Belgium Rabbit. There are two production models such as pure production and crossing-breeding production.

Different rabbit production model

Sire(♂)	Dam(♀)	Offspring	Production model	Utilization
California	California	California	Pure	Breeds
New Zealand	New Zealand	New Zealand	Pure	Breeds
Chinchilla	Chinchilla	Chinchilla	Pure	Breeds
Belgium rabbit	Belgium rabbit	Belgium rabbit	Pure	Breeds
New Zealand	California		Crossing-breeding	Fryers
California	California		Crossing-breeding	Fryers
California	New Zealand		Crossing-breeding	Fryers
Belgium rabbit	New Zealand		Crossing-breeding	Fryers

2 Reproductive Rate

Rabbits can reach sexual maturity at four months of age. However age of sexual maturity is greatly influenced by weight of the animal. In most cases, sexual maturity is reached when 75%–80% of mature body weight is achieved. It is advisable to start mating rabbits at: ① 5 to 6 months for light breeds 6 to 7 months for medium breeds. ② 9 to 12 months for heavy breeds.

Underfeeding, feeding low quality feeds and poor health can delay the age of sexual maturity, breeding management should aim for:

(1) Litter size ranging from 4 to 12 bunnies.

(2) However, litter size of 8–12 is considered commercially viable.

(3) At least 4–6 kindling per year. Bunnies should be healthy and does re-mated 3–4 weeks after giving birth.

(4) High growth rate and meat to bone ratio. This usually is function of breed.

(5) Using both breeding does and bucks for 2–3 years. Breeding can take place all year round.

3 Breeding Does and Bucks

Does, can be mated at any time after they reach sexual maturity as long as they are not pregnant. Rabbits do not have a clear reproductive cycle but the doe has periods during which they exhibit greater willingness to mate. The doe can be put with the buck at any time as the egg will come free after mating. A doe maintained in good health will ideally produce litter until they are 2.5–3 years of age.

Mature bucks can have a maximum of seven mating times per week whereas the younger growing ones can have two mating times per week.

4 Reproductive Procedure

4.1 Mating

There is no definite heat period for the non-pregnant doe and she can accept the buck at any time. When she is ready for mating the doe becomes restless, tamps its feet on the floor and will accept mounting by the buck. Mating can be considered to have taken place when the male falls aside or backwards after mounting the doe.

As Rule of Thumb When Mating Rabbits

(1) Always take the doe to the buck's cage for mating for about 30 minutes to allow several mating to take place, although a single mating can be sufficient to get the doe pregnant.

(2) Always check both rabbits for signs of infection on the genitals. Infected ones must not be bred.

(3) Do not leave the doe with the buck overnight or for a few days since this can cause abortion or females can get pregnant twice because of the y-shaped nature of their womb which can make them carry two litters at once.

(4) Re-take the doe to the buck THREE days after mating to check if the previous mating was successful. If it was, then the doe will refuse to be mounted.

(5) Do not re-mate all does that will have been confirmed to be pregnant because they may abort.

4.2　Pregnancy

The duration of pregnancy is about 30 to 33 days. Pregnancy diagnosis can be carried out by palpating the doe 12 to 14 days after mating. The procedure is:

(1) Hold the rabbit vertically with its back against the hander's body with one hand supporting the chest.

(2) Feeling the womb for the presence of the fetuses with the other

hand. With presence, the rabbit producer will be able to do this fairly easily.

(3) Increase daily amount of feeds for pregnant doe after 3th week of gestation, this benefit to the growth of baby.

(4) To ensure successful pregnancy, preparations should be done for kindling.

(5) Place a sanitized nest box in the maternity cage/ hutch during 4th week of gestation.

(6) Provide good quality dry bedding material. This could be dry grass, shredded paper or wood shavings.

(7) If the doe makes her nest on the floor, a newspaper with its edges turned up against the walls of the hutch should be place underneath the nest and litter moved into proper box just after birth.

4.3 Kindling

The litter size range from 4 to 12 bunnies, with an average of seven bunnies. A litter size of 8 to 12 is considered commercially viable. When the doe is ready to kindle, she usually plucks hair from her abdomen to line the nest. During kindling, cover the top of the nest box to avoid destructions. Young rabbits are deaf and blind and should be handled with care. Soon after kindling.

(1) Foster some of the bunnies to even out the numbers if several does kindle at the same time.

(2) Rub Vicks and foster doe's urine or feaces on handler's hands or on doe's nose to prevent rejection of bunnies by the foster doe.

(3) Allow the forested bunnies to get the first milk (colostrum) from their mother before being fostered.

(4) If the floor of the maternity cage is of wire or has big holes which make it difficult for the kindles to put the feet down, place a piece of plywood or something similar in a corner so they can rest easily .

(5) Cull the weak young rabbits within 24 hours.

(6) Remove the nest box three weeks after kindling.

Cannibalism can be caused by unrest, lack of drinking water and lack minerals or it can simply be a bad characteristic of a doe, which must be culled after does kindle ,give the doe enough water (some salt, sugar and medicine) for drinking.

4.4 Fostering

This is the remove of bunnies from a doe with a big litter size to one with a smaller one. It should be done on the first day and only for does that have kindled on the same day.

4.5 Management of the Young

The young are totally dependent on the mother's milk for the first 10 days of life. The doe can only suckle her young once or twice a day and might not be in the nest that often. After kindling:

(1) Do not disturb the doe and its young except when providing food, water and cleaning hutch.

(2) Check the nest box to remove any spoiled bedding since dirt environments can make the young susceptible to infections.

(3) Check to ensure that the bunnies are suckling and not showing any signs of sickness.

(4) Bunnies showing signs of ill health should be isolated from the rest of the litter and treated.

(5) Ensure that doe and litter and are well protected from extreme weather conditions and stress from predators.

Development stages of young rabbits: ① Day 7: body weight is doubled and fur begins to grow. ② Day 10: eyes begin to open. ③ Day 12: ears open. ④ Day 18: litter leaves nest and begins to eat solid food. ⑤ Day 21: peak milk production although litter is beginning to consume solid feed. ⑥ Day 28: milk production declines and litter is fully on solids. ⑦ Day 30–35: fully weaned. ⑧ Day 150: sexual maturity.

(6) Providing creep feeds for young rabbits.

The daily milk amount of doe is the peak of 180g on day 21, after day 12, give more feed of 100–150g per day, it can produce more milk for bunnies suckling, but after day 21, the milk of doe cannot meet the demand for bunnies growth, so we should prepare creep feed for Supplementary feeding to suckling bunnies.

On the day 12, rub the wet creep feeds into bunnies' mouth (from side in case beating your fingers) once a day, after a week training, the bunnies can learn eating the creep feeds in order to prevent some diseases, you can put some medicine in it before weaning, feeding the creep feeds and young grass for bunnies.

4.6 Weaning

Bunnies should be weaned at 30–35 days of age especially if they are low plane of nutrition. For production purpose weaning can be done as early as four weeks to ensure higher reproductive rate.

(1) When weaning, take the doe to another hutch and leave the young in their hutch. This is done to prevent stress of change of environment as young rabbits are prone to infections at this time.

(2) Ensure a clean environment at this stage for both the doe and bunnies since dirty environment can predispose the doe to mastitis.

(3) Avoid giving the doe concentrate feed 2 days after weaning to aid in drying-off if the doe was initially being fed on concentrates.

Reference Feed Formula

(1) Formula 1 for breeding rabbit .Maize 20% wheat bran 22% soybean meal 17% maize bran 38.2% Calcium hydrogen phosphate 0.3% calcium carbonate 1% salt 0.5% premix 1%.

(2) Formula 2 for bunnies between 35–90 days of age .maize 20% wheat bran 20 % soybean meal 16% maize bran 40.8% Calcium hydrogen phosphate 0.5% calcium carbonate 1.2% salt 0.5% premix %(complex vitamin 40 g soda 100 g complex enzyme 100 g choline chloride 30 g terramycin powder 30 g).

(3) Formula 3 for litter between 15–35 days of age, you can buy pellet from stock feeds factory.

Common breeds of rabbits in Zimbabwe

Breed	Breed description
California	Medium-sized breed. Predominantly white with black nose, ears, feet and tail. Very prone to skin cancer because of the white skin. Mature adult weight between 3.6–4.8 kg.
New Zealand	Medium sized breed. Predominantly white or red in colour. Early maturing, Hardy breed, Produces large litter size. Mature adult weight between 4.1–5.5 kg.

Breed	Breed description
Chinchilla	Heavy breed. Predominantly wild agouti in colour. Huge and muscular back legs, Ideal breed for meat. Mature adult weight between 5.4–7.3 kg.
Belgium rabbit	Heavy breed. Could be tan or fawn in colour, Large frame and huge size, Ideal for meat. High feed consumption coupled with high growth rate. Mature adult weight is about 5.4–5.9 kg.

Practical Technology for Raising Tilapia in Cages

1 Species Selection

The most appropriate species or strains of tilapia for cage culture are Oreochromis niloticus (Nile tilapia), O. aurea (blue tilapia), Florida red tilapia, Taiwan red tilapia, and hybrids between these species and strains. The choice of a species for culture depends mainly on availability, legal status, growth rate and cold tolerance. Many states prohibit the culture of certain species. Unfortunately, Oreochromis niloticus, which has the fastest growth rate, is frequently restricted. The ranking for growth rate of the remaining species or strains are Florida red tilapia > Taiwan red tilapia > O. aurea. Hybrids of Oreochromis niloticus × Taiwan red tilapia grow as fast as O. niloticus. Hybrids of T. aurea × Florida red tilapia grow at an intermediate rate between Florida and Taiwan red tilapia. Cold tolerance, important in Texas Agricultural Extension northerly latitudes, is greatest in O. aurea. Tilapia can be cultured at high densities in mesh cages that maintain free circulation of water. Cage culture offers several important advantages. The breeding cycle of tilapia is disrupted in cages, and therefore mixed-sex populations can be reared in cages without the problems of recruitment and stunting, which are major constraints in pond culture. Eggs fall through the cage bottom or do not develop if they are fertilized. (Reproduction will occur in cages with 1/10-inch mesh or less, which is small enough to retain eggs.) Other cage advantages include: flexibility of management; ease and low cost of harvesting; close observation of fish feeding response and health; ease and economical treatment of parasites and diseases;and relatively low capital investment compared to ponds and raceways.

2 Some Disadvantages

Risk of loss from poaching or damage to cages from predators or storms; less tolerance of fish to poor water quality; dependence on nutritionally-complete diets and greater risk of disease outbreaks; In public waters, cage culture faces many competing interests and its legal status is not well defined. Not all bodies of water offer proper conditions for cage culture.

3 Design and Construction

Both floating surface cages and standing surface cages are used for tilapia culture. Standing cages are tied to stakes driven into the bottom substrate, whereas floating cages require a flotation device to stay at the surface. Flotation can be provided by metal or plastic drums, sealed PVC pipe, or styrofoam. Cages should be constructed from materials that are durable, lightweight and inexpensive, such as galvanized and plastic coated welded wire mesh, plastic netting and nylon netting. Welded wire mesh is durable, rigid, more resistant to biological fouling, and easier to clean than flexible material, but is relatively heavy and cumbersome. Plastic netting is durable, semi-rigid, lightweight and less expensive than wire mesh. Cages made of nylon netting are not subject to the size constraints imposed by other construction materials. Nylon mesh is inexpensive, moderately durable, lightweight and easy to handle. Nylon is susceptible to damage from predators such as turtles, otters, alligators and crabs. An additional cage of larger mesh and stronger twine may be needed around nylon cages. Mesh size has a significant impact on production. Mesh sizes for tilapia cages should be at least 1/2 inch, but 3/4 inch is preferred. These mesh sizes provide adequate open space for good water circulation through the cage to renew the oxygen supply and remove waste. The use of large mesh size requires a larger fingerling size to prevent gill entanglement or escape. For example, a 3/4-inch plastic mesh will retain 9-gram tilapia fingerlings while a 1-inch mesh requires a fingerling weighing at least 25 grams with plastic netting and 50 to 70 grams with nylon netting. Larger mesh size facilitates the entry of wild fish into the cage. These fish will grow too large to swim out of the cage, but they do not grow large enough to reach marketable size, thereby representing a waste of feed. Cage size may vary from 1 to more than 1,000 cubic meters. As cage size increases, costs per unit volume decrease, but production per unit volume also decreases, resulting from a reduction in the rate of water exchange. Cages should be equipped with covers to prevent fish losses from jumping or bird predation. Covers are often eliminated on large nylon cages if the top edges of the cage walls are supported 1 to 2 feet above the water surface. Feeding rings are usually used in smaller cages to retain floating feed and prevent wastage. The rings consist of small-mesh (1/8 inch or less) screens suspended to a depth of 18 inches or more. Feeding rings should enclose only a portion of the surface area because rings surrounding the entire cage perimeter may reduce water movement through the cage. However, feeding rings that are too small will allow the more aggressive fish to control access to the feed. If sinking feed is used, small cages may require a feed tray to minimize loss. These rectangular trays can be made of galvanized sheet metal or mesh (1/8 inch; galvanized or plastic) and are suspended from the cover to a depth of 6 to 18 inches.

4 Site Selection and Placement of Cages

Large bodies of water tend to be better suited for cage culture than small ponds because the water quality is generally more stable and affected less by fish waste. Exceptions are eutrophic waters rich in nutrients and organic matter. Small (1 to 5 acre) ponds can be used for cage culture, but provisions for water exchange or emergency aeration may be required. Cages should be placed where water currents are greatest, usually to the windward side. Calm, stagnant areas should be avoided. However, areas with rough water and strong currents also present problems. Cages may be moored individually or linked in groups to piers, rafts, or lines of heavy rope suspended across the water surface. At least 15 feet should separate each cage to optimize water quality. The cage floor should be a minimum of 3 feet above the bottom substrate, where waste accumulates and oxygen levels may be depressed. However, greater depths promote rapid growth and reduce the possibility of parasitism and disease. See SRAC publications Nos. 160–166 for more information on cage culture.

5 Production Management

Geographic range for tilapia culture is temperature dependent. Preferred water temperature range for optimum growth is 82 to 86 F. Growth diminishes significantly at temperatures below 68F and death will occur below 50 F. Only the southernmost states have suitable temperatures to produce tilapia in cages. In the southern region tilapia can be held in cages from 5 to 12 months per year depending on location.

6 Fingerlings

Cages may be used for fingerling production. One-gram fry may be reared in 1/4-inch mesh cages at up to 3,000 fish per cubic meter for 7 to 8 weeks until they average about 10 grams each. Ten-gram fish can be restocked into 1/2-inch mesh cages. Cages stocked with 10-gram fish at 2,500 per cubic meter will produce 25- to 30-gram fingerlings in 5 to 6 weeks. After grading, 25- to 30-gram fish can be restocked at 1,500 fish per cubic meter to produce 50- to 60-gram fingerlings in 5 weeks, or at 1,000 fish per cubic meter to produce 100-gram fingerlings in 9 to 10 weeks. Fish should be graded by size every 4 to 6 weeks. Stunted fish and females should be culled.

7 Final Growout

The optimum fingerling size for stocking in final growout cages is determined by the length of the growing season and the desired market size. The shorter the growing season, the larger the fingerlings must be at stocking. The use of male populations which grow at twice the rate of female populations will result in larger fish, greater production and a reduction in the growout period. In temperate regions, overwintered, 1-year-old fingerlings of 60 to 100 grams (4 to 7 fish/pound) are generally used to produce fish of 1 pound or greater in cages. If 1/2-

pound fish are acceptable for market, then it maybe possible to rear smaller, 20- to 30-gram fingerlings (15 to 23 fish/pound) which were produced during the spring of the same year. Recommended stocking rate of tilapia fingerlings depends on cage volume, desired harvest size and production level, and the length of the culture period. Expected harvest weights of male tilapia are given in Table 1. High stocking rates can be used in small cages of 1 to 4 cubic meters. Optimum stocking rates per cubic meter range from 600 to 800 fish to produce fish averaging 1/2 pound; 300 to 400 to produce fish averaging 1 pound; and 200 to 250 to produce fish averaging 1.5 pounds. Water exchange is less frequent in large cages, and therefore the stocking rate must be reduced accordingly. In 100-cubic meter cages, the optimum stocking rate is approximately 50 fish per cubic meter to produce 1-pound fish. In temperate regions, complete or batch harvests are required. Cages for final growout should be stocked when water temperature rises above 70°F and harvested when the temperature falls below 70°F. In tropical or sub-tropical regions with a year-round growing season, a staggered production system could be used to facilitate marketing by ensuring regular harvests, e.g., weekly, biweekly, or monthly. The exact strategy will depend on the number of cages available and the total production potential of the body of water. Example: if 10 cages are available for placement in a pond with sufficient production potential and grow out takes 20 weeks, then a cage could be stocked every 2 weeks. Beginning on week 20, the first cage would be harvested and restocked, followed by another cage every 2 weeks. A staggered system requires a regular supply of fingerlings.

The Prevention and Control Knowledge of Zoonotic Diseases Preventive Methods in Zimbabwe

1 Concept of Zoonosis

Zoonoses are diseases caused by the same pathogen, which are epidemiologically related and naturally spread between animals and humans.

According to statistics, there are more than 500 kinds of pathogenic organisms harmful to animals and people, animals and people will have more than 100 infections in their lifetime, but the infections that cause disease are usually only 5 to 10 times, this is mainly animal and human disease resistance barrier (skin, mucous membrane, leukocyte, antibody) play a role.

2 Classification of Zoonosis

There are many kinds of zoonoses, which are classified according to certain rules, in this way easy to understand, study, control and eliminate the zoonoses. There are many methods to classifying the zoonoses, also including the academic categories, zoonoses can also be classified according to the need for prevention and control. Basically, the criteria can be determined and classified, according to the pathogen, epidemic link, distribution range, prevention and control strategies of the zoonoses.

Based on the traditional methods, the 90 zoonotic diseases further divided into 14 categories, according to the characteristics of the pathogens.

(1) DAN viral zoonosis. Cowpox, monkeypox, sheep mouth sore, etc.

(2) RNA viral zoonosis. Influenza, rabies.

(3) Zoonosis caused by the prion virus. Scrapie, mad cow disease, etc.

(4) Zoonosis of gram-negative bacteria. Brucellosis, plague, etc.

(5) Zoonosis of gram-positive bacteria. Anthrax, etc.

(6) Actinomycosis. Tuberculosis and so on.

(7) Rickettsia. Ornithosis, Q fever, etc.

(8) Spirochetes. Leptospirosis.

(9) Mycosis. Tinea, etc.

(10) Protozoosis. Toxoplasmosis, etc.

(11) Paragonimiasis. Schistosomiasis, lung fluke.

(12) Tapeworm. Cysticercosis, hydatidosis of sheep, etc.

(13) Nematode. Trichinella, ascaris, etc.

(14) Infestation ectoparasites.Ticks, mites, maggots, lice, fleas.

3 Hazards of the Zoonotic Diseases

Zoonotic diseases not only prevalent in ancient times and modern times, but also the biggest killer of human beings of today, humans are still unable to fully control the occurrence and prevalence of the zoonoses. Zoonoses cause not only massive human death, disability and incapacity to work, it has brought about biological disasters, economic hardship to many families and seriously affected the social stability.

It's estimated that 17 million people die from infectious diseases every year all over the world, 95% are in developing countries, among them the main infectious diseases are zoonoses, tuberculosis alone kills 3.1

million people each year and more than 10,000 people die from rabies. Brucellosis is very serious in developing countries, it has high levels of infection and cause serious health damage for people and animals.

The periodic prevalence of the zoonoses , attracted the most concern at the worldwide, people still not ignore the rabies, tuberculosis, influenza, plague and so on, and implementing the early warning and prevention methods.

4　Diagnosis of Zoonosis

(1) Sample collection. Specimens should be collected timely and correctly according to the route of pathogen invasion and the regular pattern of development, in this way that is easy to diagnose and determine.

(2) For the parasitic diseases, the autopsy, microscopic examination of insect eggs, smear, and other techniques can be adopted.

(3) Bacteriosis can be detected by adopting the staining microscopic examination, separation and purification, classification, and animal experimental testing techniques.

(4) Viral diseases can be adopted the testing methods of the isolation, antibodies neutralize, PCR and other technologies.

(5) Allergic reaction. It uses T-lymphocytes from animals, to be sensitized by antigens, using antigenic substances for allergy testing, for example tuberculin eye dotting , intradermal allergic reaction.

(6) Antigens and antibodies react directly, that is able to use the visible response of complete antigens and antibodies for diagnosis. Such as agglutination reaction, precipitation reaction, AGAR diffusion, neutralization experiment.

(7) Antigen or antibody is labeled for diagnosis, there is no visible

reaction to antigens and antibodies, therefor which is displayed by red blood cells, enzymes and fluorescence. Since antibodies appear only at certain stages in the course of the disease or during recovery, serological methods are not very effective in early diagnosis. There are undetected error, at the same time, also have false positive and false negative problems occur due to cross reactions. These are concepts of the sensitivity, specificity, true positive rate and false positive rate in serological diagnosis.

5 Epidemiological Investigation

The basic knowledge of the epidemiological investigation, can get by the basic knowledge of the various zoonotic diseases. According to this, there are four main aspects:

5.1 Clinical Symptoms

Many zoonoses have typical clinical symptoms, such as neurotic symptoms of rabies and aggressive behavior. The septicemic appearance of anthrax that causes bleeding from the natural openings.

5.2 The Invasion Route of the Disease

Disease occurs in any region or any group of animals all have the invasion routes. There for, it is necessary to analyze the biological and mechanical transmission of the routes.

5.3 The Epidemiological Characteristics

The five epidemic characteristics of zoonosis should be analyzed and compared.

5.4 The Outbreak Surveillance Results

After the years of epidemic occurrence, the surveillance results are based on correct diagnosis then summarize the rules of epidemic occurrence. Through the monitoring results, the disease occurrence risk type analyzed to determine the possibility of disease occurrence.

6 Protection Requirements for Diagnosis of Zoonotic Diseases

The zoonotic diagnoses are a dangerous jobs, this jobs may go deep into the epidemic area, contact with the pathogens, it may become infected and spread the pathogens. Therefore, the protection work should be including the four aspects.

(1) Carefully implementing of the personal protection. Staff should wear protective clothes and pay attention to the disinfection in case of suspected severe zoonosis.

(2) Follow the proper operating procedures. The treatment of the animals with unknown causes of death should be distinguished from the general and the special circumstances. The identification of the epidemic should focus on the basis of comprehensive analysis, not blindly necropsy of the dead animals.

(3) Operate according to the technical specifications. The disease materials, diagnosis and other operations, should be carry out in the strict requirements of the relevant technical specifications .

(4) Strengthen the laboratory biosafety management.

Prevention and Treatment of Disease Animals

1 Tick-borne Disease

1.1 Diagnosis and Treatment

In general, specific laboratory tests are not available to rapidly diagnose tick-borne diseases. Due totheir seriousness, antibiotic treatment is often justified based on clinical presentation alone.

1.2 Lyme Disease or Borreliosis

Symptoms: Fever, arthritis, neuroborreliosis, erythema migrans, cranial nerve palsy, carditis, fatigue, and influenza-like illness; Treatment: Antibiotics-amoxicillin in pregnant adults and children, (doxycycline in other adults).

1.3 Relapsing Fever (Tick-borne Relapsing Fever, Different from Lyme Disease Due to Different Borrelia Species and Ticks)

Organisms: Borrelia species such as B. *hermsii, B. parkeri, B. duttoni, B. miyamotoi*; Treatment: Antibiotics are the treatment for relapsing fever, with doxycycline, tetracycline, or erythromycin being the treatment of choice.

1.4 Typhus Several Diseases Caused by Rickettsia Bacteria (below)

Symptoms:Fever,headache, altered mental satus, myalgia,and rash; Treatment: Antibiotic therapy, typically consisting of doxycycline or tetracycline.

1.5 Helvetica Spotted Fever

Symptoms: Most often small red spots, other symptoms are fever, muscle pain, headache and respiratory problems; Treatment: Broad-spectrum antibiotic therapy is needed, phenoxymethylpenicillin likely is sufficient.

1.6 Human Granulocytic Anaplasmosis (Formerly Human Granulocytic Ehrlichiosis or HGE)

Organism: Anaplasmaphagocytophilum (formerly Ehrlichia phago-cytophilum or Ehrlichia equi); Vector: Lone star tick (Amblyomma americanum), I.scapularis; Bartonella: Bartonella transmission rates to humans via tick bite are not well established but Bartonella is common in ticks.

2 Disinfection of Chicken Farms

2.1 Dry Cleaning

Sweep or blow dust and other loose dirt off ceilings, light fixtures, walls, cages or nest boxes, fans, air inlets etc. onto the floor. Remove all feed from feeders. Scrape manure and accumulated dust and dirt from perches and roosts. Remove all litter from the floor. Litter can be added to a compost pile. Sweep the floor to remove as much dry material as possible. With a small coop, a wet-dry shop vacuum does a good job of removing this material. However, be careful to clean the filter often as the fine dust from the coop may easily clog the filter and make the vacuum work harder or lead to burn out of the motor.

2.2 Wet Cleaning

Turn the power off to the building prior to using any water for cleaning. Wet cleaning is done in three steps: soaking, washing and rinsing. Warm or hot water will do a better job getting through organic matter than cold water. You can use a cheap neutral detergent, like dish soap.

Soaking

Soak the heavily soiled areas (perches and roosting areas, floors, etc.) thoroughly. Use a low pressure sprayer to totally soak all surfaces. Soak until the accumulated dirt and manure has softened to the point it is easily removed.

2.3 Washing

Wash every surface in the building, especially window sills, ceiling trusses, wall sills and any surface where dirt and dust may accumulate. The washing solution can be either a neutral detergent (ph between 6 and 8) or an alkaline detergent (ph above 8). Alkaline substances vary in their strength with the strongest causing burns and internal injuries if swallowed. A mild alkali is baking soda (sodium bicarbonate) and moderate alkalis include household ammonia, borax and trisodium phosphate. Strong alkalis include washing soda (sodium carbonate) and lye (caustic soda). Mix in hot water—160°F or hotter is best.

2.4 Disinfecting

This is a crucial step which the small flock owner might normally overlook. Disinfectants should be applied only after the building and equipment have been thoroughly cleaned, ideally right after rinsing. Disinfectants can be applied by sprays, aerosols or fumigation. Don't be intimidated by the thought of "fumigating" your hen house: for most small flock facilities, using a garden type sprayer is the easiest method, and chances are you already have a suitable disinfectant around the house. The types of disinfectants generally used are phenolic compounds (e.g., Pine-sol, One Stroke,), iodine or iodophors, (e.g., Betadine and Weladol), chlorine compounds (e.g., Clorox, generic bleach), quaternary ammonium compound (e.g., Roccal D Plus) and oxidizing compounds (e.g., Virkon S, Oxy-Sept 333).

Follow the manufacturer's directions for mixing and dilution of these disinfectants. A good rule of thumb is to apply at the rate of one gallon of diluted disinfectant per 150–200 square feet of surface area. For a more thorough disinfecting, soak water and feeders in a 200 mg/kg chlorine solution (1 tablespoon chlorine bleach per gallon of boiling water).

3 Mastitis in Dairy Cattle

3.1 Management

Cattle affected by mastitis can be detected by examining the udder for inflammation and swelling, or by observing the consistency of the milk, which will often develop clots or change color when a cow is infected.

Another method of detection is the California mastitis test, which is designed to measure the milk's somatic cell count as a means for detecting inflammation and infection of the udder.

3.2 Treatment

Treatment is possible with long-acting antibiotics, but milk from such cows is not marketable until drug residues have left the cow's system. Antibiotics may be systemic (injected into the body), or they may be forced upwards into the teat through the teat canal (intramammary infusion). Cows being treated may be marked with tape to alert dairy workers, and their milk is syphoned off and discarded. To determine whether the levels of antibiotic residuals are within regulatory requirements, special tests exist. Vaccinations for mastitis are available, but as they only reduce the severity of the condition, and cannot prevent reoccurring infections, they should be used in conjunction with a mastitis prevention program.

3.3 Control

Practices such as good nutrition, proper milking hygiene, and the culling of chronically infected cows can help. Ensuring that cows have clean, dry bedding decreases the risk of infection and transmission. Dairy workers should wear rubber gloves while milking, and machines should be cleaned regularly to decrease the incidence of transmission.

3.4 Prevention

A good milking routine is vital. This usually consists of applying a pre-milking teat dip or spray, such as an iodine spray, and wiping teats dry prior to milking. The milking machine is then applied. After milking, the teats can be cleaned again to remove any growth medium for bacteria. A post milking product such as iodine-propylene glycol dip is used as a disinfectant and a barrier between the open teat and the bacteria in the air. Mastitis can occur after milking because the teat holes close after 15 minutes if the animal sits in a dirty place with feces and urine.

Handbook on Agricultural Machinery Maintenance Technicals Diagnosis and Elimination of Common Tractor Faults

1 Clutch Faults

Clutch common faults are clutch slip, clutch separation is not complete, shaking, abnormal sound, etc. The diagnosis and troubleshooting of the fault are shown in Tables below, respectively.

Diagnosis and elimination of the fault of clutch slip

Cause of fault	Elimination method
• Clutch pedal free travel is too small. • Clutch separation system failure. • Failure of compression mechanism. • Friction plate damaged, worn, oil polluted.	• Check and adjust the pedal free travel. • Check whether the clutch pedal stuck, whether the return is strong, the three clutch separation lever height is in line with the requirements, then repair. • Check the clutch cover and flywheel fastening screws, if loose, to fasten, if pressure spring failure, to replace, clean the polluted oil on pressure plate. • Check the friction sheet, flywheel end surface have polluted oil or not, friction sheet is becoming thin, deformation or not.

Diagnosis and troubleshooting of incomplete clutch separation

Cause of fault	Elimination method
• Clutch free clearance is too large. • Clutch separation system failure.	• Check and adjust the clutch pedal free travel. • Check and adjust the three separation lever height and separation lever bracket, bracket pin.

Cause of fault	Elimination method
• The friction sheet warps. • The engine is not fixed firmly or the crankshaft axial clearance is too large.	• The forward gear and reverse gear tests are carried out respectively. If the feeling is heavy and changing, the failure will be driven plate warping, loose rivet or friction plate breaking. • If you don't feel stable, maybe the engine is not fixed firmly or crankshaft axial clearance is too large.

Diagnosis and elimination of clutch shaking

Cause of fault	Elimination method
• Pedal free travel is too small. • The separation lever height is different. • The separation fork fulcrum is worn. • The driven friction sheet, damping spring damaged. • Transmission, flywheel housing, engine not fixed well.	• Check and adjust the pedal free travel. • Check to adjust the separation lever so that the inner end faces are on the same plane. • Check the wear of the separation fork fulcrum and repair it. • Replace the driven friction plate, damping spring. • Fix well the front and rear bracket, transmission and flywheel of engine.

Diagnosis and troubleshooting of abnormal clutch sound

Cause of fault	Elimination method
• Pedal return spring damaged. • Separation bearings lack of oil or damage. • Pressure plate fits clutch cover loosely.	• Idle operation abnormal sound disappeared when the pedal will be activated, indicating that the pedal return spring damaged, then replace. • Lightly step on the clutch, just to eliminate the free travel rustling sound, grease added. Still ring after filling, then step down a little and slightly speed up, the noise increases, indicating that the separation bearing is damaged and replaced. • Steping pedal deeply causes the abnormal sound, which increases with the increase in speed, stable operation at medium speed, sound significantly reduced or disappeared, pedal released, sound disappeared, the clutch is normally working, this is the pressure plate and clutch cover with loose, then repair.

Cause of fault	Elimination method
• Separation lever or bracket pin loose, friction rivets loose, exposed. • Driven disc and spline rivet loose or steel plate broken. • Spline shaft sleeve fits the first shaft loosely, the spring broken.	• Continuously step down the clutch pedal, in the union and separation of the moment there is a strange sound, it is loosely that the separation of lever or bracket pin or friction rivets loose, exposed, then repair. • When starting, metal dry friction sound and accompanied by shaking phenomenon, the fault is driven disk and spline rivet loosely or steel sheet broken, then repair. • If in the combination of a collision sound, generally spline shaft sleeve and the first shaft with loose, repair. If the driven plate with a shock absorber, it may be the shock spring broken, replace.

2 Transmission Faults

Due to the poor driving conditions of tractors, transmission failures are frequent. The failure should be promptly removed, so as not to cause more damage to the parts. The common transmission failure phenomenon, the reason and the elimination method are shown in Tables below.

Diagnosis and elimination of shifting gear difficultly or not shifted

Cause of fault	Elimination method
• Clutch separation is not complete. • Improper adjustment of variable speed interlock mechanism. • The spline of transmission shaft is worn or cracked. • Fork end wear, fork loose, deformation or broken. • Fork shaft positioning ring groove wear. • The spring force on the lock pin is too large to cause the lock pin to be stuck.	• Adjust the clutch separation clearance. • Adjust variable speed interlock mechanism. • Overhaul or replacement. • Maintenance fork. • Repair fork shaft ring groove or replace. • Adjust the spring force on the lock pin.

Diagnosis and elimination of gearbox abnormal sound and automatic off-gear fault

Phenomenon of fault	Cause of fault	Elimination method
Gearbox abnormal sound	• Gearbox lack of oil. • The gear wears out and the gap becomes larger. • Spline or spline hole wear, the gap becomes larger. • The gearbox shaft locking device loose, free movement of the shaft.	• Add oil. • Check or replace the gear and adjust the clearance. • Check or replace splines or spline shafts or gears, and adjust the clearance. • The locking device of the shaft to be fastened.
Automatic off-gear	• The lock pin spring of variable speed shaft is broken or invalid. • The gear face is badly worn or meshing is too short. • The fork is deformed.	• Replace the spring. • Replace the gear and adjust the meshing stroke. • Correct or replace the fork.

3 Universal Joint and Drive Axle Failure

Common faults of universal joint and drive axle include abnormal sound of drive shaft, abnormal sound of drive axle and heat of drive axle, etc. The phenomena, causes and troubleshooting methods are shown in Tables below.

Diagnosis and elimination of abnormal sound fault of drive shaft

Phenomenon of fault	Cause of fault	Elimination method
The abnormal sound only appears above the medium speed, and the higher the speed, the greater the sound, when reaching a certain speed, the body vibration and shaking, at this time off the gear sliding, vibration and shaking more intense.	• Universal joint journal wear, needle wear or damage. • The drive shaft is bent. • The balance weight falls off or the assembly mark is out of alignment. • The coaxiality of the cross axis and the drive shaft is poor.	• Repair universal joint journal and needle roller. • Overhaul the drive shaft. • Check the balance weight and align with the assembly mark. • Abnormal sound for the periodic snoring, put up the rear axle to shift high speed gear and neutral gear to check. If the abnormal sound is a continuous vibration, idle running and observe the swing of the drive shaft. Check the coaxiality of the cross shaft and the drive shaft.

Diagnosis and elimination of abnormal sound of drive axle

Phenomenon of fault	Cause of fault	Elimination method
Abnormal noise of rear axle when driving. Straight driving no abnormal sound, abnormal sound when turning. Abnormal noise when going uphill or downhill. Rear wheels have running noise or heavy abnormal noise.	• The main reducer oil is insufficient or bad. • Gear meshing is bad. • Bearings and gears are worn or damaged. • Spline, thrust gasket, cross shaft wear. • The tooth of the gear was broken. • There is a fault in the differential.	• According to the regulations to add or replace the main reducer gear oil. • Adjust the meshing gap of the gears. • Maintenance of bearings and gears (the higher the speed, the louder the sound. Failure in bearing wear, too small meshing gap or too large gear end wear). • Maintenance spline, thrust gasket, cross shaft (variable speed abnormal sound, fault in the meshing gap is too large, half shaft and spline meshing loosely). • Change the gear (the abnormal sound will disappear when the gear is out of gear and sliding, indicating that the teeth of the main driven gear are broken). • Check differential (abnormal sound when turning, the fault is differential).

Diagnosis and elimination of heating fault of drive axle

Cause of fault	Elimination method
• Insufficient lubricating oil of the main reducer. • The drive and driven gear meshing gap is too small. • Differential bearing and driving gear bearing pretension over large.	• Add lubricating oil according to the regulations. • Adjust the meshing gap between the drive and driven gears. • Adjust the preload of the bearing.

4 Steering and Walking Systems Faults

The common fault phenomena, causes and troubleshooting methods of steering system and walking system are shown in Tables below.

Diagnosis and troubleshooting of heavy steering faults

Cause of fault	Elimination method
• Frame, front axle, suspension deformation, front wheel positioning misalignment, tire pressure is too low, wheel valley bearing is too tight. • Steering gear lack of oil, with too small clearance, bad meshing, steering shaft bending. • Steering transmission mechanism is lack of oil, assembly is too tight, bearing damage, steering rod, steering knuckle arm deformation.	• Jacking up the front bridge, if the steering wheel feel easy, then the fault in the frame, axle, wheels, suspension, check whether the front wheel positioning is normal and tire pressure is too low, wheel damage bearings are too tight, front axle, leaf spring and frame with deformation or not. • Feel heavy then remove the steering arm, if the steering is still heavy then the fault is in the steering gear, check in and eliminate lack of oil of the steering gear, fit gap is too small, bad meshing, steering shaft bending and other problems. • When the steering arm is removed, it is easy to feel the fault in the steering transmission mechanism. Check and rule out the problems such as lack of oil or assembly of the steering transmission mechanism, bearing ring damage, deformation of the steering rod and knuckle arm, etc.

Diagnosis and elimination of driving deviation

Cause of fault	Elimination method
• The front tire pressure is inconsistent. • The front tire is out of alignment. • Front wheel hub bearings are not elastic. • Brake drag. • Leaf springs broken, riding bolts loose, resulting in too much spring. • Frame, front axle deformation, left and right wheelbase difference is too big. • Knuckle arm, knuckle deformation or loose.	• Check and adjust the front tire pressure and make them consistent. • Adjust the front wheel beam. • Overhaul the front wheel hub bearing. • Eliminate the brake drag. • Check whether the leaf spring is broken, whether the riding bolt is loose and rule out. • Overhaul the frame, front axle and adjust the wheelbase. • Overhaul steering knuckle arm and knuckle.

Diagnosis and elimination of driving swing fault

Cause of fault	Elimination method
• The installation and adjustment of the steering gear is not in place. • Steering gear wear, front hub bearing axial clearance is too large, front wheel positioning misalignment. • Frame, suspension and other deformation.	• Maintenance steering gear part: if the steering wheel free travel is too large, indicating that the steering gear meshing transmission pair clearance is too large. If the steering wheel is loose, the worm or screw upper and lower bearings are adjusted too loose or the steering assembly is installed loose. Should be adjusted or tightened. Check and adjust the clearance between the steering boom shaft and the bushing. • Overhaul the steering transmission part: the two front wheels are facing the front, turn the steering wheel, observe whether the ball pin of each tie rod is loose. Support the front axle to check wheel hub bearing clearance, kingpin and bushing clearance, etc. If abnormal wear is found in the front tire, check the front tire band. • Repair the car frame, suspension, etc.

5 Brake System Fault

The brake system fault will seriously affect the safety of the tractor, so it should be checked and eliminated in time. Common failure phenomena, causes and troubleshooting methods are shown in Tables below.

Diagnosis and troubleshooting of brake failure

Phenomena of fault	Cause of fault	Elimination method
• The tractor can not slow down or stop when the brake pedal is pressed down or a foot brake is pressed to the bottom.	• The brake main pump is short of oil. • Brake tubing rupture or joint oil leakage. • Part of the mechanical connection fell off. • The main pump ring damage or aging.	• Continuously step down the brake pedal can not step in the end, should first check whether the main pump is short of oil, and add oil according to the regulations. • If not short of oil, check front and rear the brake tubing is broken or oil leakage, and exclude. • Check whether the transmission rod parts fall off. • Finally check the main pump ring is damaged or aging, and exclude.

Diagnosis and troubleshooting of bad brake

Phenomena of fault	Cause of fault	Elimination method
Poor braking effect, too long braking distance.	• The main pump oil hole plug, oil valve damage, have air in the system. • The brake pedal free travel is too large. • Brake clearance is too large, friction disc serious wear or bad contact. • Brake pump stuck valve. • Brake fluid is missing, there is air in the pipeline. • There is a leak in the brake piping system.	• Exhaust air in the brake pipe, if the brake is still bad, check the brake main pump. • Check and adjust the brake pedal free travel. • Adjust the gap between the friction disc and the brake drum, clean and wash or replace the friction disc. • Clean the brake pump. • Add brake fluid and remove the air in the pipeline. • Eliminate the leak.

Diagnosis and elimination of brake deviation

Phenomena of fault	Cause of fault	Elimination method
When braking, both sides of braking effect is not the same.	• The brake gap of one side wheel is too large, so that the brake shoe cannot press the brake drum, and the friction decreases. • The brake disc of one side wheel is badly worn or damaged. • There is dirty oil on one side of the brake or the sub pump damaged. • Brake pump balance valve failure or throttle valve seal. • The two rear tire pressure is inconsistent. • Left or right brake pipe air intake.	• Check the side of the brake gap of the poor brake effect, do not conform to the specified value to adjust. • Replace the brake friction disc, etc. • Remove the side of the wheel brake of the poor braking effect, clean up the dirty oil. • Replace the parts. • pump up a tire at the prescribed pressure. • Exhaust air.

Diagnosis and elimination of brake drag fault

Phenomena of fault	Cause of fault	Elimination method
Tractors are difficult to start and difficult to run. After pressing the clutch pedal, the vehicle speed decreases obviously. The brake drum heats up after a certain distance.	• Brake pedal free travel is too small. • The gap between friction disc and brake drum is too small. • The main pump leather ring swelling, back to the oil hole blocked. • Pump ring swelling or piston back blocked.	• Adjust the brake pedal free travel. • Clearance between friction disc and brake drum . • After adjusting stroke and normal clearance, check the main pump leather ring and the same spring, the same oil hole. • The main pump is normal then check sub pump: jacking up the body, turn the wheels in turn, the fault is on the big resistance wheel with the idling.

Mechanization Technology of Saving Cost and Increasing Efficiency in Wheat Production

1 Introduction

(1) Mechanization technology in wheat production results in increased efficiency as well as saving of seed, fertilizer, chemicals, water, fuel, labour among other things.

(2) Therefore, it is necessary to use high-tech agricultural machinery in every stage of wheat production.

(3) Use of high-tech agricultural machinery in every stage of wheat production promotes the incorporation of straw into the field, deep fertilizer application, precise sowing, energy-saving plant protection, water-saving irrigation, use of combine harvesters and other cost-saving and efficient mechanization technologies.

(4) This is the main technical measure that assures an increase in efficiency, income and food security.

2 Incorporation of Corn Straw into the Field

(1) This technology mainly uses the corn combine harvester and the straw pulverizing machine to form a complete set.

(2) After harvesting, the straw is pulverized and incorporated into the soil.

(3) This is very important because when the straw/ crop residue decomposes, it becomes organic fertilizer which improves soil fertility

thereby enhancing wheat production.

(4) Widespread application of this technology, not only improves the utilization of resources, but also avoids burning of straw hence reducing environmental pollution and other related problems.

(5) This also reduces labour requirements by 80–100 times.

(6) According to the experimental measurement, the direct incorporation of fresh corn straw into the field is equivalent to applying standard nitrogen fertilizer of 300–450 kg per hectare, standard phosphate fertilizer of 150–300 kg and potash fertilizer of 450–650 kg.

(7) The soil organic matter content increases by 0.1% or so. This can increase the wheat yield by 8%–12%.

3 Deep Chemical Fertilizer Application Mechanization Technology

(1) The mechanization technology of deep application of fertilizer is such that in the process of tillage and wheat planting, fertilizer is applied precisely in right quantities and to the required depth to allow ease absorption by plants.

(2) The results showed that the utilization rate of chemical fertilizer could be increased by 10%–15%, the fertilizer efficiency could reach more than 46%, and the grain yield per hectare could be increased by 300–600 kg.

(3) At the same time, it reduces the environmental pollution caused by chemical fertilizer volatilization and improves the economic and social benefits of farmers.

(4) This technology mainly includes deep application of basal fertilizer, deep application of seed fertilizer and deep application of top

fertilizer.

(5) Deep application of bottom fertilizer, mainly using the combined operation of arable land fertilization machine.

(6) Fertilizer is evenly applied into the bottom of the plough soil layer.

(7) Fertilizer is applied to a depth of 10–25 cm, to achieve the same operation of tillage and fertilization.

(8) Deep application of seed fertilizer, mainly using grain fertilizer seeder, sowing and at the same time applying fertilizer, covering and pressing.

(9) This application is to a depth of 3–5 cm.

(10) Deep application of topdressing, that is, in the process of wheat growth, according to the agronomic requirements of the topdressing timing and quantity, the use of topdressing machine or tillage fertilizer machine, the fertilizer will be applied to the lower part of the wheat side, generally 8–12 cm, and to a depth of 6–10 cm, coverage rate reaches 100%. This also ensures there is adequate compaction.

(11) Precision sowing is a high-yield wheat cultivation technology, with large and medium-sized wheeled tractors and matching wheat precision planters.

(12) The outstanding characteristic is that seed is dropped in right quantities, correct depth and at uniform spacing.

(13) Normally 1–3 seeds per every 5 centimeters length. The depth is usually 3–5 cm.

(14) Due to planting of seed in correct quantities there is reduction of amount of seed planted per hectare from 105–135 kg to 45–90 kg. This will lead to about 52.5 kg of seed per hectare being saved.

(15) Due to uniform sowing, the plant nutrition, ventilation and exposure to sunlight can be improved.

(16) More effective tillers lead to a well-developed root system and high growth rate thereby resulting in 8%–13% increase in yield compared to conventional sowing.

4 Energy-saving Plant Protection Mechanization Technology

(1) Powder spraying and ultra-low volume spraying are two common techniques of energy-saving in plant protection. Powder spraying technology has the characteristics of good atomization performance, long range, high efficiency, wide adaptability and so on.

(2) Under normal circumstances, the dosage of 450 liters per hectare is called conventional spray, 45–450 liters is called low-dose spray, and the dosage below 45 liters is called ultra-low-dose spray.

(3) If Ultra-low quantity spraying technology is adopted, the mist flow is fine, the concentration is high, the drift in the crop cluster is good, the penetration is good, the control effect is good.

(4) This can save the pesticide by 10%–30% compared with the conventional spraying. Therefore, it should be mainly promoted and applied.

5 Water-saving Irrigation Mechanization Technology

(1) Water-saving irrigation mechanization technology is the main technical measure adopted to improve the utilization rate of water resources and achieve high yield and efficiency of crops, including spray technology, drip irrigation technology, seepage prevention technology of channels and low-pressure pipeline water transfer technology.

(2) the popularization and application of these technologies, the water can be saved by 30%–50%, and the wheat yield can be increased by 6%–9%.

6 Combined Wheat Harvesting Mechanization Technology

(1) Combined wheat harvesting mechanization technology can be

used to complete the wheat harvesting, threshing, cleaning, transporting and other operations.

(2) Compared with manual segmenting harvesting,

the working efficiency is improved by 160-220 times, and the harvest period is shortened by 5-7 days.

(3) At the same time, it can reduce the physical labour of farmers and liberate a large number of labour force to invest in other industries.

(4) At present, the market sales of machinery and tools are mainly piggyback and self-propelled.

(5) The self-propelled is divided into large, medium and small models.

(6) The quality of operation and use performance are very good.

(7) When harvesting the wheat there is need to bring enough disassembly and assembly tools, spare parts and wearing parts, in order to timely troubleshoot and improve the utilization rate of the machinery.

Rotor Tiller

1 Preparation

(1) Dress properly: Sturdy foot wear, Work suit.

(2) Inspect the area to be tilled: Remove foreign objects, Do not till above water lines, electric cables or any pipe networks.

(3) Disengage all clutches, all control levers must be in neutral before starting the engine.

(4) Fuel handling: Use an appropriate container, never add fuel to a running engine or hot engine, fill fuel tank outdoors., replace gasoline/fuel cap securely and clean any spillage, never attempt to make adjustments while the engine is running.

2 Operation

(1) Never operate the tiller without guards, covers and hoods in place.

(2) Never start the engine or operate the tiller with wheels in the free-wheel position. Make sure the wheel lock pins are engaged through the wheel hubs and wheel axle.

(3) The wheels act as a brake to keep the tiller at a controlled speed.

(4) Disengage wheel lock pins to permit free wheel only when engine is stopped.

(5) Keep hands, feet and clothing away from rotating parts.

(6) Tines and wheels rotate when tiller is engaged in forward or reverse.

(7) In forward, tines and wheels rotate when the drive safety control lever(Forward), is pushed down towards the handlebar.

(8) In reverse, wheels and tines rotate when the drive safety control lever(reverse), is pulled back towards the operator.

(9) Releasing the reverse handle to the neutral position stops the wheels and tines.

(10) Do not operate both drive safety control levers at the same time.

(11) Be extremely cautious when operating in reverse.

(12) After striking a foreign object, stop the engine, remove the wire from the spark plug and inspect the tiller for any damage and repair the damage before restarting and operating the tiller.

(13) If vegetation clogs the tines, raise the handlebars to elevate the

tines and run the tiller in reverse. If this does not clean the vegetation from the tines stop the engine and disconnect the spark plug wire before removing vegetation by hand.

(14) Do not touch the engine muffler with bare hands to avoid severe burns since it will be hot during operation.

(15) If the unit vibrates abnormally, stop the engine and check the cause.

(16) Do not run the engine indoors because the exhaust fumes are deadly.

(17) Do not overload the machine capacity by tilling too deep at a too fast rate.

(18) Never operate the machine at high speeds on slippery surfaces.

(19) Never allow bystanders near the unit.

(20) Use attachments and accessories approved by the manufacturer of the tiller.

(21) Never operate the tiller without good visibility or light.

(22) Be careful when tilling in hard ground, tines may get stuck in

the ground and propel the tiller backward. If this occurs, let go of the handlebars and do not restrain the machine.

(23) Take all possible precautions when leaving the machine unattended. Disengage all control levers, stop the engine, wait for all moving parts to stop, and make sure that guards and shields are in place.

3 Maintenance and Storage

(1) Keep machine, attachments and accessories in safe working condition.

(2) Check shear bolts engine mounting bolts and other bolts at frequent intervals for proper tightness to be sure the equipment is in safe working condition.

(3) To prevent accidental starting always disconnect and secure the spark plug wire from the spark plug before performing tiller maintenance.

(4) Never run the engine indoors. Exhaust fumes are deadly.

(5) Always allow muffler to cool before filling fuel tank.

(6) Never store equipment with gasoline in the tank inside a closed building where fumes may reach an open flame or spark. Allow the engine to cool before storing in any building.

(7) Always refer to the operator's guide instructions for important details if the tiller is to be stored for an extended period.